THE SAMUEL & ALTHEA STROUM
LECTURES IN JEWISH STUDIES

SAMUEL Stroum, businessman, community leader, and philanthropist, by a major gift to the Jewish Federation of Greater Seattle, established the Samuel and Althea Stroum Philanthropic Fund.

In recognition of Mr. and Mrs. Stroum's deep interest in Jewish history and culture, the Board of Directors of the Jewish Federation of Greater Seattle, in cooperation with the Jewish Studies Program at the University of Washington, established an annual lectureship at the University of Washington known as the Samuel and Althea Stroum Lectureship in Jewish Studies. This lectureship makes it possible to bring to the area outstanding scholars and interpreters of Jewish thought, thus promoting a deeper understanding of Jewish history, religion, and culture. Such understanding can lead to an enhanced appreciation of the Jewish contributions to the historical and cultural traditions that have shaped the American nation.

The terms of the gift also provide for the publication from time to time of the lectures or other appropriate materials resulting from or related to the lectures.

A Best-Selling Hebrew Book of the Modern Era

THE *BOOK OF THE COVENANT* OF PINḤAS HURWITZ AND ITS REMARKABLE LEGACY

David B. Ruderman

A Samuel and Althea Stroum Book

UNIVERSITY OF WASHINGTON PRESS
Seattle and London

This book is published with the assistance of a grant from
the Samuel and Althea Stroum Endowed Book Fund.

Printed and bound in the United States of America
Composed in Minion Pro, typeface designed by Robert Slimbach
18 17 16 15 14 5 4 3 2 1

UNIVERSITY OF WASHINGTON PRESS
www.washington.edu/uwpress

LIBRARY OF CONGRESS CATALOGING-IN-PUBLICATION DATA
Ruderman, David B.
A best-selling Hebrew book of the modern era : the book of the covenant of Pinḥas Hurwitz and its remarkable legacy / David B. Ruderman. — 1st edition.
p. cm. — (The Samuel & Althea Stroum lectures in Jewish studies)
Includes bibliographical references and index.
ISBN 978-0-295-99413-0 (hardcover : alk. paper)
1. Hurwitz, Phinehas Elijah, 1765–1821—Sefer ha-berit. 2. Science. 3. Judaism and science. I. Title.
Q157.H79R833 2015
500—dc23 2014007528

The paper used in this publication is acid-free and meets the minimum requirements of American National Standard for Information Sciences—Permanence of Paper for Printed Library Materials, ANSI Z39.48–1984.∞

To my daughter-in-law, Dr. Asha Ravikumar,
and to my son-in-law, Rabbi Joshua Strom

Contents

Preface

DEVOTING an entire monograph to the subject of a single relatively unknown Hebrew writer of the eighteenth century, Pinḥas Hurwitz, and his best-selling *Sefer ha-Brit* (Book of the Covenant) requires some explanation. I have been aware of this composite Hebrew work—part scientific encyclopedia, part ethical guide, and part mystical ascent—for several decades now. My own long fascination with it actually began in the 1980s when I was studying Jewish culture in early modern Italy. In the aftermath of Francis Yates's provocative book on Giordano Bruno and her controversial understanding of the "Hermetic" origins of modern science,[1] I composed a book on the sixteenth-century Italian Jewish thinker Abraham Yagel and the bizarre mixture of kabbalah, magic, and science in his writings.[2] Already then, I noticed a book first published in 1797, some two hundred years later, grappling with similar but seemingly belated issues. Given my own limitations as an early modernist, Hurwitz and his intellectual world appeared too remote in time and place to consider then as I tried to understand Yagel's complex constructions of knowledge. But because Hurwitz's book seemed so far from Baroque Italy, it intrigued me all the more. It was clearly a book with strong lines of continuity with the past, especially one surprisingly rooted in Italy, but it was clearly not an early modern creation, even if one were to extend the artificial boundaries of this period to the end of the eighteenth century. It was a modern book, and its author was a modern author, not a throwback to an earlier age or to an earlier mindset. In finally returning to Hurwitz and his intellectual world, I wish to explore why this modern author and his readers still appear to be preoccupied with issues primarily shaped in a previous age.

Since I wrote my earlier books some twenty-five years have passed.[3] Scholarship on Jewish history and on early modern and modern European thought has evolved in multiple ways, as has my own thinking about the early modern era and its continuities and discontinuities with the epoch that followed it. The earlier focus of a previous generation on a canon of German Jewish thinkers from Mendelssohn to Buber and Rosenzweig has shifted to a greater recognition of the importance of eastern European culture and thought, to the flowering of rabbinic creativity, and to a consideration of secondary figures who were significant carriers of culture within the Jewish community. My attempt to reconstruct Pinḥas Hurwitz, his remarkable book, and its enormous impact on modern Jewry might certainly be linked to some of these new trends in scholarly writing, particularly work that has focused on the Gaon of Vilna and his students, the *Haskalah* (Jewish enlightenment) and its opponents, the history of Ḥasidism in its larger social and political contexts, the lingering impact of Sabbateanism and Frankism on European Jewry, and the history of modern orthodoxy, to name only a few significant areas of contemporary research.[4]

Sefer ha-Brit's importance as a Jewish book of the late eighteenth century lies in its resistance to simple classification: neither modern nor antimodern; neither religious nor secular; and neither orthodox nor maskilic. Because it could not easily be labeled under any one ideological banner, it was widely read, and it stimulated Jewish readers from every ideological and religious sector of Jewish society. On the one hand, it opened to thousands of readers an enticing taste and appreciation of the natural world, modern science and technology, and the thrill of new discoveries and inventions. It also offered a strong social critique of Jewish society and its attitudes toward others, both Jew and non-Jew. On the other hand, it led its readers on a religious and mystical journey, inspired by none other than the kabbalistic teaching of Ḥayyim Vital, the mystic of Safed (1543–1620), and his quest for holiness, along with the spiritual testimonies of other Jewish thinkers, precursors and disciples of Isaac Luria and the pietistic explosion he generated for centuries.

To understand Hurwitz's book is to appreciate its composite nature and to recognize how it resonated among so many segments of modern Jewish society both in the West and in the East. While Hurwitz claimed to have come from Vilna and spent most of his later years in Cracow, it would be too simple to label him an eastern European Jew, presenting a different cultural alternative to Jews of the West. In fact, he was transnational and transregional, an eastern European Jew westernized by his travels to Berlin, the Hague, and London and transformed by his encounter with books, authors, and acquaintances he acquired during his travels.

Hurwitz was not only an author; he was also a highly successful book dealer. He is a modern author both because of what he wrote and because of how he promoted his work. The grafting of kabbalah and science had a long history before Hurwitz, but no one before him had created so appealing a fusion, which made Hurwitz one of the most popular of modern Hebrew writers. And no one before him had succeeded so well in mastering his craft and the business of printing, which allowed him to produce his book while creating an impressive network of sales representatives to hawk it aggressively. Hurwitz might be connected to a long line of writers originating during the Renaissance, if not before, but he singularly discovered the formula of financial success within the eighteenth-century marketplace of books he knew firsthand.

That remarkable combination of intellectual curiosity, social critique, moralistic fervor, intense spirituality, and business acumen accurately define the author of *Sefer ha-Brit* as a modern Jewish author. And conversely, the amazing success of the book among modern readers defies the notion that it is merely a belated product of a previous era. The book was read and appreciated by booklovers well into the twentieth century; it was a standard reference work that rested on the shelves of many Jewish households, and it was considered a book that reflected the tastes and interests of modern consumers and students of Hebrew books. Given its popularity, it requires us to ponder more deeply its comfortable and natural place within modern Jewish culture. What did modern readers find so appealing about the book that they did not find, in dramatic contrast, among earlier authors writing about kabbalah and science or among Hurwitz's own contemporaries writing Hebrew compositions on science? Neither of these authors sold anywhere near the number of books vended by Pinḥas Hurwitz either during his lifetime or well after.

The book before the reader offers a series of preliminary inquiries into the central themes of *Sefer ha-Brit*. It represents an initial probe of a complex work that has never been studied systematically and thoroughly. As a prefatory assessment of Hurwitz's intellectual world and readership, I am fully aware of its shortcomings. Hurwitz's book, first and foremost, lacks a critical edition, based on a full accounting of the extensive changes and additions between the first and second enlarged edition, as well as careful comparison of the texts of each subsequent edition throughout the nineteenth and twentieth centuries. Because such a task is beyond my abilities, I have relied on the most accessible and complete text now available, the *Sefer ha-Brit ha-Shalem* published in Jerusalem in 1989–90, though I have compared it on occasion with the first edition and with other earlier editions.

While I think I have gained a general sense of the written sources Hurwitz utilized in composing his work, my assessment is neither exhaustive nor definitive. In chapter 3 below, I discuss briefly his sources on science and natural philosophy, Jewish and non-Jewish, known to me based on the work of others or on my own investigations. These sources, though preliminary, constitute, I suspect, a sufficient sampling to substantiate the conclusions I draw below. I also have compiled a list of most of the references to Hebrew sources that Hurwitz used in writing his book, some of which I refer to in subsequent chapters. But I am sure I have missed several, especially those he does not mention by name, including those from which he silently copied. While it is obvious how influential Lurianic kabbalah, especially the writings of Ḥayyim Vital and later expositors of Lurianic kabbalah on Italian soil, was on Hurwitz's kabbalistic orientation, I cannot investigate this subject with the authority and thoroughness I would like. I will leave this task to an expert in kabbalistic literature of the early modern period.

The book opens with a hypothetical meeting in The Hague, which might have taken place but actually never did, between Naphtali Ulman, a local Jewish philosopher and student of Leibniz and Wolff, and the young visitor to the Dutch city, Pinḥas Hurwitz. This allows me to introduce Hurwitz by way of a study in contrast, charting the transition in the second half of the eighteenth century between Ulman's understanding of Judaism, philosophy, and science and that of Hurwitz. Having introduced Hurwitz in chapter 1, I then reconstruct his life and career as thoroughly as I can in chapter 2, with special attention to the production and distribution of his book and his appeal to his many readers.

Chapter 3 treats his attitude toward the natural world and raises in particular the question of how he conjoined his preoccupation with nature with his devotion to the kabbalah. Chapter 4 examines his attitude toward philosophy. While clearly objecting to the dominant place of philosophical metaphysics in Judaism, Hurwitz creatively used both philosophical and historical arguments to make his case. Of special importance is his usage of Kant in justifying his merger of natural philosophy and kabbalistic metaphysics.

In chapter 5, I treat Hurwitz's most original discourse, on the love of all human beings, Jews and non-Jews alike, and attempt to contextualize his radical conclusions within Jewish and non-Jewish thought at the end of the eighteenth century. Focusing on this large chapter where few sources are mentioned by the author, I tentatively suggest a masonic background to explain Hurwitz's thinking, based especially on his description of a well-known hero of Freemasonry.

The last chapter presents a wide panorama of the readers of *Sefer ha-Brit*, beginning with two contemporaneous critics and ending with several positive endorsements from the end of the twentieth and beginning of the twenty-first centuries. In presenting his diverse readership—including Maskilim (the proponents of the *Haskalah*), Ḥasidim, and their opponents the Mitnagdim, ashkenazic and sephardic rabbis, and modern Yiddish and Hebrew writers—I demonstrate that Hurwitz's boastful declaration, in the second introduction of *Sefer ha-Brit*, that his book was a major success was hardly an empty claim. In fact, the author could never have imagined how well the book actually sold for well over two hundred years after his death.

I close this book with an epilogue that offers some final thoughts on *Sefer ha-Brit* and its place in modern Jewish culture. I contextualize my findings on Hurwitz by reference to three recent works on modern Jewish culture, which I present briefly. Three appendices follow: a list of all the known editions of the book; a full translation of Hurwitz's unusual printing instructions for the book, discussed in chapter 2; and a simple table of contents of *Sefer ha-Brit*.

Acknowledgments

I AM especially indebted to several colleagues who offered me their constant support, encouragement, and specialized knowledge as I undertook the writing of a book that often stretched the limits of my own competence in Jewish history. They include Yisrael Bartal, Eliezer Brodt, Richard I. Cohen, Resianne Fontaine, Moshe Idel, Irit Idelson-Shein, Maoz Kahana, Rachel Manekin, Jonatan Meir, Yitshak Melamed, Elhanan Reiner, Asher Salah, and Eliyahu Stern. My sincere thanks to them all.

I was fortunate to have receptive academic audiences to try out my initial reflections on *Sefer ha-Brit* as they slowly emerged. I want to thank especially the following institutions that invited me to speak on various aspects of the book: Kings College, London; the University of Antwerp; Heinrich Heine University, Düsseldorf; Hochschule für Jüdische Studien, Heidelberg; Goethe University, Frankfurt am Main; the American Academy in Berlin; Martin Luther University, Halle-Wittenberg; Ludwig Maximilian University, Munich; the Centro de Ciencias Humanas y Sociales, Madrid; Central European University in Budapest; Princeton University; the Graduate Center of the City University of New York; Rutgers University; Boston University; the Material Texts Seminar of the University of Pennsylvania; Johns Hopkins University; Ohio University; the University of Arizona; and Tel Aviv University.

An earlier version of this book was presented as the Samuel and Althea Stroum Lectures in Jewish Studies at the University of Washington, Seattle, on October 22–26, 2012. I am most grateful to Professor Noam Pianko; his assistant director, Lauren Spokane; and the other faculty members of the

Stroum Center for their kind invitation to present these lectures and for their hospitality. I have enjoyed working with the pleasant and efficient editorial staff of the University of Washington Press, especially its director, Nicole F. Mitchell, assistant acquisitions editor Tim Zimmermann, David Myers, and Jonathan Meir. I am thankful for Dave Peattie at BookMatters, and I have also benefited greatly from my highly skilled copyeditor, Anne Canright.

Some parts of the book were earlier published or will appear at about the same time as this book's publication. Chapter 1 was first published as "The Hague Dialogues" in *Mapping Jewish Amsterdam: The Early Modern Perspective—Dedicated to Yosef Kaplan on the Occasion of His Retirement*, ed. Shlomo Berger, Emile Schrijver, and Irene Zweip, *Studia Rosenthaliana*, vol. 44 (Leuven: Peeters, 2012), 221–239. A part of chapter 2 appears as "On the Production and Dissemination of a Hebrew Best Seller: Pinḥas Hurwitz and His Mystical-Scientific Encyclopedia, *Sefer Ha-Brit*," in a Festschrift honoring Anthony Grafton, edited by Ann Blair, Anja Goeing, and Urs Leu (forthcoming from Leiden: Brill). A small section of chapter 5 serves as a basis for the expanded Hebrew essay "The Mental Image of Two Cherubim in Pinḥas Hurwitz's *Sefer ha-Brit*: Some Conjectures," in a Festschrift honoring Richard Cohen, edited by Ezra Mendelsohn (forthcoming from Jerusalem: Merkaz Shazar).

As always, I benefited from the extraordinary support of my colleagues at the Herbert D. Katz Center for Advanced Judaic Studies, who allowed me the time to write this book. I want to thank especially Sheila Allen, Carrie Love, and Etty Lassman, as well the helpful library staff of the Katz Center library, especially Judith Leifer. In completing this book at the end of my last year (my twentieth) as director of the Center, I offer my sincere appreciation to the entire staff, the numerous fellows I have met, and the board members of this remarkable institution.

My wife, Phyllis, offered me her usual loving support and confidence in me. In dedicating this book to my new "children," Dr. Asha Ravikumar, my daughter-in-law, and Rabbi Joshua Strom, my son-in-law, I wish to express my love and affection for both of them.

A BEST-SELLING HEBREW BOOK
OF THE MODERN ERA

CHAPTER 1

The Hague Dialogues

IMAGINE the following scenario: A young scholar, apparently from Vilna, having wandered through several cities in eastern Europe and Germany, arrived in the city of The Hague toward the close of the 1780s. Once there, he most likely enjoyed the material support of the richest family of Jewish merchants in the city, the Boas family, and sought and gained the religious approval of the rabbi of the city, Judah Leib Mezerich. The young man's name was Pinḥas Elijah ben Meir Hurwitz (1765–1821), and he was about to complete the first draft of his soon-to-be published book, an encyclopedia of the sciences entitled *Sefer ha-Brit* (Book of the Covenant).[1] Hurwitz soon learned of the presence of an aging sage who lived in the city, a rigorous philosopher and émigré from Mainz, Naphtali Herz Ulman (1731–87). Ulman had completed a multivolume philosophic opus of which only the first volume, *Ḥokhmat ha-Shorashim* (The Science of Roots or First Principles), had been published, in 1781.[2] Hurwitz himself was hardly a philosopher; rather, he was a student of religious texts, drawn especially to the kabbalah. But he did share something in common with Ulman: an appreciation of the life of the mind and particularly a fascination for the natural world and the new sciences. They were also both ashkenazic Jews with knowledge of the German language.[3] It seems natural that Hurwitz would seek out Ulman, the major intellectual figure of Hague Jewry.

While this scenario may have been possible, it was probably never realized. Ulman, who had lived in Holland for more than fifteen years, died in 1787 at the age of fifty-six.[4] Hurwitz, according to the testimony of Mezerich, was living in The Hague in 1790 and had been residing there for at least a full year.[5] If indeed he arrived in 1789, he had missed his opportunity to engage with the formidable philosopher. If only they had had a chance to meet, one might have imagined an animated, even contentious conversation. Ulman was thirty years older than Hurwitz, and that generational difference in age revealed an enormous gap in their intellectual styles and in the values and aspirations they held for themselves and their communities. *Sefer ha-Brit* was first published in 1797 in Brünn, only sixteen years after Ulman's single Hebrew publication, but the contrast between the two works—not to mention the numerous volumes Ulman left in manuscript but never published—is astounding. If one were to chart the transformation of Jewish thought from one generation to another, these two thinkers would provide meaningful markers of some continuity but more of radical change. Ulman's cultural world was that of Leibniz, Wolff, and Mendelssohn, the ambiance of the German enlightenment, with which he identified even from his self-imposed exile in Holland. Hurwitz was of a different makeup. While he knew of Mendelssohn and his philosophical contemporaries, he had chosen a different intellectual trajectory, merging his passion for Lurianic kabbalah with an appreciation and commitment to the study of the natural world. Not Wolff nor Leibniz but Kant inspired him. Even Maimonides, the quintessential medieval Jewish philosopher, excited him far less than his self-proclaimed spiritual mentors Isaac Luria and Ḥayyim Vital of Safed. Judging from the intellectual products of these two Hebraic scholars who nearly encountered each other in the Jewish neighborhood of The Hague, the interval of a mere sixteen years had utterly altered the intellectual landscape of the Jewish cultural world.

Historians should not engage in hypotheticals regarding what might have happened had two historical figures encountered each other. But in the case of Ulman and Hurwitz, two scholars who happened to reside in the same place at nearly the same time, their literary legacies offer ample information to reconstruct what a conversation between them might have sounded like. Far removed from the cultural centers of Berlin, Cracow, and Vilna, The Hague might seem an unlikely cultural setting for monitoring a critical transition in modern Jewish thought, an evolution from Wolffian metaphysics to kabbalistic natural science. Neither Ulman nor Hurwitz ever attained a status as primary intellectual figures in the history of modern Jewish thought, a designation that would certainly include the likes of Mendelssohn, Maimon, and other luminaries of the German *Haskalah*. But they remain fascinating secondary figures of Jewish self-reflection, and they help to illuminate major shifts in the cultural world of European Jewry at the end of the eighteenth and beginning of the nineteenth centuries—a transitional period usually charted through Mendelssohn, his children and disciples and adversaries, primarily in Germany. It also might properly set the stage to consider the content and impact of Hurwitz's best-selling book on several generations of Jewish readers to follow. This opening chapter, therefore, which I call "The Hague Dialogues," depicts a conversation that, although it most likely never took place, seems quite plausible in the light of the extensive literary remains of these two thinkers. It will set the scene for the book that follows.

THE NEED FOR RABBINIC APPROBATIONS

We might begin our comparison of Ulman and his younger contemporary, Hurwitz, by perusing the openings of their published volumes. The title page of *Ḥokhmat ha-Shorashim*, published by two local printers, Loeb ben Moses Soesmans and J. H. Munnikhuisen, with the financial support of four local Jewish patrons,[6] indicates immediately that wisdom of the roots of knowledge represents the first science of metaphysics and serves as an introduction to the volumes that follow: wisdom of the world, the soul, and finally the divine. Having prepared volumes on each of these topics as well as other works, Ulman obviously imagined that this initial publication of 1781 would be followed by more. Perhaps one reason for the book's lack of financial success and his failure to publish any subsequent volumes can be found in this statement of the author:

> I apologize in publishing this book for having abandoned the custom of most writers of our times in approaching the famous, brilliant, and wise rabbis who sit in learned counsel, teachers of the Torah in the land of Holland where I presently reside, to seek their permission and approval of this publication. I justify this on two grounds. First, something that is made clear by decisive proof, which is the case for all the studies discussed in this book, is complete and self-evident truth and requires no formal approval. Would it not be ridiculous, for example, to declare that the three angles of a triangle represent more than two sides of a perpendicular just because a famous person gave his approval of this statement? Second, these rabbis, who are famed throughout the land as our teachers and sages, are so burdened with Torah and divine commandments and with their profound and

> expert investigations of heavenly law and religious observance that they have no spare time to contemplate the wonders of nature and the secrets of the divine through self-reflection. The only demonstrative proof for them is revealed tradition, as it is for the majority of loyal and faithful Jews.[7]

Ulman's sarcastic and belligerent stance toward the rabbis of his generation in justifying his decision to ignore any rabbinic *haskamah* (approbation) for his book could not have endeared him to them and their loyal constituencies, the potential purchasers of his Hebrew tome. On the contrary, he was both rejecting their authority in understanding the truth and also claiming that the study of nature and metaphysics had no place in their own curriculum of sacred study. In this he openly declared that the only authentic expositors of the truth were philosophers like himself (note his own self-designation on the title page: "Torah scholar, naturalist [*mehandes*], and exalted philosopher"),[8] who alone were capable of leading the Jewish community in these confused and troubling times.

How different was Hurwitz's strategy in *Sefer ha-Brit*! It appears that Hurwitz went to great lengths to solicit *haskamot* from seven rabbis during the early 1790s: Saul Loewenstamm, rabbi of the ashkenazic community of Amsterdam; David Azevado, rabbi of Amsterdam's sephardic community; Aryeh Leib Breslau, rabbi of Rotterdam; the aforementioned Rabbi Judah Leib Mezerich of The Hague; Isaac ha-Levi of Lemberg, rabbi of Cracow; Moses Münz of Brod, rabbi of Oven (Ofen = Buda); and Rabbi Isaac Abraham of Pintshov (Pinczow). Hurwitz apparently lobbied hard for these rabbinic approbations. When approaching Rabbi Loewenstamm of Amsterdam, for example, he brought with him a letter of introduction from the latter's own brother, Rabbi Ẓevi Hirsch of Berlin. Rabbi Azevado surely knew of Rabbi Loewenstamm's support for Hurwitz's book, as he mentioned that he was writing his own approbation during the mourning period for this recently deceased colleague, who had just died on June 19, 1790. Rabbi Mezerech had been ordained by Rabbi Loewenstamm and was surely inclined to follow in his footsteps in writing his own approbation. Rabbi Isaac ha-Levi of Cracow was the son-in-law of Rabbi Aryeh Leib Breslau of Rotterdam and was surely in contact with him. In short, Hurwitz chose his seven rabbis carefully. Despite the distances from eastern Europe to Holland, these rabbis were connected to each other and, at least in some cases, simply followed the lead of their associates in granting rabbinic approval of Hurwitz's book.[9]

Hurwitz's motivation in enlisting the support of these rabbinic authorities was surely related to their privileged place within traditional Jewish society. In contrast to Ulman, who consciously maligned these figures by making fun of their authority and knowledge, Hurwitz knew full well that he needed *haskamot* to sell his book among traditional Jews. Furthermore, he assumed that if these rabbis would issue a warning prohibiting anyone from republishing the book for fifteen years without the permission of the author, he would be relatively protected to pursue his own publishing interests and would reap any profits exclusively for himself. To his utter surprise, this expectation was not realized. Despite the rabbinic threats mentioned in the approbations, a pirated edition of *Sefer ha-Brit* appeared in 1800–1801, and the publisher ignored the rabbis altogether, going so far as to remove their approbations from his edition. They were restored in Hurwitz's new and expanded edition of 1806–7, which included a new introduction in which Hurwitz expressed his disgust over the sheer disregard by the rogue publisher in ignoring these rabbis who had supported his publishing endeavor.[10] In short, in the mind of Hurwitz rabbinic approbations were still important, both in providing religious approval of his message and in protecting him from illicit publishing practices that would harm his authorial rights. When this proved not to be the case, Hurwitz not only excoriated the man who had stolen his book; he also singled out this practice at the end of *Sefer ha-Brit* in a broader social critique of the Jewish community of his day.[11]

ON METAPHYSICS

In the same year that Ulman published his introduction to Jewish metaphysics, or what he called "The Wisdom of That Which Is Beyond Nature,"[12] Immanuel Kant published his *Critique of Pure Reason*.[13] Ironically, at the very moment that Ulman was conceiving his major project of a Jewish philosophy based on the principles of Leibniz and Wolff, Kant had initiated a direct assault on those same thinkers' core assumptions. Although Ulman deviated from his mentors in some of his specific formulations, he was clearly indebted to their core ideas and the overall structures of their works, viewing his enterprise as a specific Jewish adaptation of their vital contributions to philosophical thinking. Even a superficial perusal of his ambitious composition reveals employment of their working hypotheses such as the principles of contradiction, sufficient reason, and the best and most perfect worlds, as well as a description and defense of the existence of monads.[14]

Leaving aside the intricacies of Ulman's specific interpretations of Leibniz and Wolff, I wish to stress in this context Ulman's unwavering commitment to the enterprise of metaphysical thinking in general. Right from the beginning of *Ḥokhmat ha-Shorashim*, he underscored the need for philosophical reflection in determining first principles and in organizing human knowledge. He was certainly aware of those detractors of philosophical metaphysics from within both the traditionalist and the rational camps and aggressively sought to deflect their criticisms. For Ulman, metaphysics was fundamental to human understanding; faulty principles yield bad knowledge and ultimately harm society. Ulman readily conceded that Platonic and Aristotelian philosophy had made invalid assumptions; nevertheless, the search for meaning beyond and behind nature was still valid. Metaphysical thinking was neither impractical nor dry. Without a systematic undergirding of philosophical principles, human society was cast into doubt and heresy, and for Jews, the meaning of the Torah was put in jeopardy. What the Jewish community required, Ulman believed, was a new paradigm of philosophical metaphysics based on Leibniz and Wolff and a moral commitment to fathom the ontological basis of reality rigorously and sincerely. Sensory knowledge alone could not sufficiently make sense of the infinite variety of things in nature; fundamental rational assumptions were necessary as well.[15]

Hurwitz would have strongly objected to this line of thinking. He would have particularly found offensive Ulman's declaration at the beginning of *Ḥokhmat ha-Shorashim* that anyone who assumes it is impossible to know truth except through prophecy and kabbalah is completely mistaken.[16] The proofs of "science" undermine such claims. For this indeed was the exact position that Hurwitz came to champion. He saw the kabbalah as the ultimate source of all truth and the foundation of the Jewish faith, and he viewed his primary intellectual role as an expositor of the kabbalistic tradition. In *Sefer ha-Brit,* therefore, he undertook to interpret kabbalistic sapience for a wide Jewish readership. For Hurwitz, philosophy, both in the past and in his own time, generally led to skepticism and heresy, denying the foundations of Jewish faith. He believed, rather, that it was only through faith, and not rational investigation, that a Jew could know God: "It is not God's desire . . . that we will know the Lord our God through human investigation and acquired proof . . . for God wants from his people that they believe in him based on our tradition from our forefathers, generation after generation back to those who stood on Mount Sinai."[17]

In radical opposition to the stance of Ulman, Hurwitz declared that the entire philosophical enterprise was antithetical to Judaism. Philosophy was the creation of the Greeks, beginning with Solon, Socrates, and Aristotle. The Arabs then took this tradition and transmitted it to Christian Europe. Building on the Moslems and the Christians, certain Jews mastered philosophy and wrapped themselves in "a stolen tallit."[18] Despite the attraction of philosophy for some Jewish thinkers, most notably Maimonides, philosophical study was never incumbent upon Jews and did not, Hurwitz believed, lead to knowledge of God. The latter could only be reached, following the formulation of Judah ha-Levi, by the observance of the divine commandments and by experiencing the world to its fullest. Wisdom was not a function of rational investigation but rather of faith and understanding and appreciating the natural world, the wondrous creations of the Divine: "A thing is more certain acquired through achievement and experience than one acquired through human investigation as the elder sages declare: 'There is no wiser person than one of experience.'"[19]

In almost revelatory terms, Hurwitz dramatically announced the publication of Kant's devastating critique of metaphysics in 1781. In Hurwitz's words, the book appeared to undermine not only ancient philosophy but also "the most recent philosophers of this generation . . . and their theoretical proofs regarding the reality beyond nature called metaphysics." Hurwitz explicitly pointed to "the books of the great philosopher Wolff and the famous philosopher Leibniz who are the most publicized recent philosophers," describing their intellectual systems as children's toys made of paper and carton easily blown away in the wind.[20] For Hurwitz, Kant's challenge was devastating to metaphysics, showing that all philosophical proofs are mere figments of thought and imagination unconnected to reality. Kant demonstrated unequivocally that human beings lack the means to know, based on either the senses or rational investigation, what is beyond nature.

Hurwitz added that the attempt of the contemporary Jewish philosopher Solomon Maimon to defend the foundations of philosophy was unsuccessful. Hurwitz associated Maimon's defense with a justification of Leibniz's theory of monads, which, he said, have no basis in reality and were invented in Leibniz's mind. Hurwitz acknowledged the many followers of Leibniz but considered Kant's disbelief as the ultimate vindication that such metaphysical assumptions were nothing but deceptive vanities. With the shattering of the truth claims of the school of Leibniz and Wolff, the door was open to embrace a metaphysics based on kabbalistic assumptions constituting the true faith of Judaism, one philosophically informed by Kant's latest critique and aligned with an empiricist appreciation of the physical world known through human experience.[21]

The explicit reference by Hurwitz to Kant's challenge to Leibniz and Wolff underscores the gulf that existed between Ulman and Hurwitz. While Ulman suffered the fate of defending an epistemological system already under siege by the time he had completed most of his writing, Hurwitz greatly benefited from the timeliness of Kant's challenge to metaphysics and his attempt to limit human investigation to the natural world. Through Kant, Hurwitz discovered a convincing strategy of combining natural philosophy with kabbalistic metaphysics. In this vital conflict over the place of philosophy in Judaism, Hurwitz had found a remarkably effective ally from within the very cohorts of recent philosophy from which he intended to distance himself.

ON INTERPRETING THE WONDERS OF NATURE

In one of his many compositions left in manuscript, Naphtali Ulman described an incredible invention that he had seen in the home of an affluent Jew from Amsterdam named Jacob Heimfeld. The invention was an elaborate clock, which he called a horologe; from it a sculpted figure would emerge every half hour, accompanied by music. The music, Ulman said, was so accurate that it was indistinguishable from that of a live musician. He also reported on other such clocks in which figures of stone or wood emerged on the hour, hitting a chime with a stick to indicate the precise time; with each appearance a door would open and then shut as the figure entered and exited.[22]

Ulman remarked that such an incredible sight could easily be perceived as a miracle of nature, but in fact it represented a daily natural occurrence, indeed, one that was encountered with growing regularity. In fact, he observed, such phenomena were also recorded regularly in history books and testified to the fact that what many people perceive to be miracles are actually matters of natural causation. Ulman was aware of those who would challenge this position by claiming that biblical miracles were often intended to punish the wicked and reward the righteous and that nature alone never distinguishes between right and wrong or privileges one group of people over another. But this view, Ulman argued, could be refuted by several historical examples. In 1349, the Jews were less victimized by the black plague than Christians, but this was clearly not the result of a miracle. At other times, women perished more than men in times of war. One can also point to examples of unusual natural phenomena in the Bible that are often perceived to be miraculous, such as the crossing of the sea of reeds, Sarah's pregnancy in old age, the rapid growth of Israel's population in Egypt, and more. Such events do not happen often, Ulman said, and when they do occur they often astonish human observers, but they are still natural in origin. Just because one is unfamiliar with such phenomena does not mean they should be labeled miracles. Scripture does not record irrational and illogical things. A good natural philosopher aware of the limitless possibilities within nature can ultimately explain and interpret all of the so-called miracles recorded therein.[23]

Ulman's strong claim was directed against the traditionalist rabbis of his generation who assumed that every naturalist was a nonbeliever and that the only strategy for challenging those who denied the miraculous was to retreat into a biblical literalism and fundamentalism. Ulman offered an alternative by which to reconcile God, providence, and nature. The naturalist does not deny miracles, he maintained; he simply tries to understand them within a naturalistic context. The comet was often perceived as a miracle by those who had not studied the laws of astronomy, but the scientist knew that they are part of nature. Philosophers like him were actually true believers, as opposed to charlatans within the Jewish community, on the one hand, who deceived and manipulated the masses with their pretentious claims to be miracle workers, and to deist and atheist philosophers, on the other, who denied the veracity of Holy Scripture.[24]

The true hero was what Ulman called "the Torah philosopher," who acknowledges revelation about nature but tries to offer a rational hypothesis as close to the truth as possible. In this respect, Ulman mentioned the well-publicized debate between Pierre Bayle and Gottfried Leibniz. The former had firmly distinguished between the truth of nature and the truth of scripture, thus erecting an epistemological firewall that legitimated all faith claims. Ulman, in contrast, siding with his mentor Leibniz, argued that reason and faith ultimately do not contradict each other; rather, they complement each other, since God is the source of both. The implications for the Jewish community are clear: "Every wise Jew must attempt to understand through truthful reasoning all the biblical passages in the Holy Books, to interpret them so they do not appear irrational and accordingly seem possible and truthful; or at least it is his obligation to exert himself to understand from his intelligence that there is no real contradiction embedded in them."[25]

Proving that the Bible is a repository of natural phenomena, that miracles can be explained rationally, and that the role of "the Torah philosopher" is to reconcile the scientific and the sacred were objectives that Hurwitz would have found disconcerting. To the casual observer of Hurwitz's massive encyclopedia of the sciences, which offered a comprehensive view of astronomy, the earth sciences, and human biology, this might appear odd. Did Hurwitz not promote his project of pious science as a means of integrating knowledge of the natural world with divine revelation? Why would he object to Ulman's idea that all the narratives of the Torah demonstrate the regularity and rationality of nature and support a scientific understanding of divine creation? But indeed, Hurwitz's motivation in teaching science to pious Jews was not identical to that of Ulman.

In the first place, for Hurwitz there was only one source of truth, and that was the revelation at Sinai, and specifically its deeper meaning as revealed in the kabbalah. Human beings unaided by divine revelation are in no position of knowing the truth. This is especially the case for metaphysical knowledge, as we have seen, but it is also true for knowledge of the natural world. Natural philosophers could certainly provide a general orientation on how to understand the universe and God's creation, according to Hurwitz, but their understanding was incomplete and tentative. They could never comprehend all the secrets of nature, which are revealed over time and constantly challenge the rational paradigms that philosophers use in explaining the natural world.

What was the ultimate meaning of recent scientific discoveries such as the helium balloon, the barometer, or the diving bell that Hurwitz presented to his readers in the pages of *Sefer ha-Brit*? In one sense, they convey a sense of wonderment and inspiration at the majestic beneficence of God's presence in the world, echoing the biblical notions of how the heavens declare God's glory or how from human flesh, God is revealed. But they accomplish something else as well for Hurwitz: they destabilize the orderly system of human knowledge based on long-standing rational assumptions and intense scrutiny of nature. In the chance way in which nature operates, and in its seeming indifference to the supposed rules and norms by which it is expected to function according to human calculation, it becomes apparent that human knowledge cannot reliably fathom God's handiwork.

At several points in his text, Hurwitz pauses to consider the utter confusion that scientific discoveries often render. One example was the seemingly unwavering assumption that the earth was composed of four elements, which chemical philosophers had recently challenged, insisting that there were in fact five. It was not that the later scientists were smarter than the earlier ones, Hurwitz claimed, only

> that experience has come to deny their words and falsify their opinions. And what has happened to the ancients will also happen to the scholars of our generations so that the moderns will ultimately not be happy. Then in the end of days, discoveries will be revealed which are still hidden from us through the invention of new instruments, and these will contradict the assumptions of the scholars of our own generation. Those future scholars will no longer remember favorably the views of our contemporary ones, for in another generation all they accepted will be erased. Their sons will rise after them to discover even better instruments and clearer experiments to deny their former conclusions. Everything they had accepted will be destroyed by propositions based on philosophical logic resting on the wings of rational human assumptions. . . . Thus all external sciences are far from the truth with the exception of mathematics and geometry. Everything else will be swept away in a wind of vanity, since it has no force to establish the truth.[26]

In contrast with Ulman's emphatic assumption that philosophers can establish a relatively stable and reliable system of metaphysical and physical truth that can be made compatible with the utterances of the Torah, Hurwitz saw the study of nature from an entirely different vantage point. The more one learns of nature and the ways in which scientists work, he said, the more one realizes how tentative and fallible human knowledge and experience are. By revealing the limitations of human understanding, empiricism in fact bolsters the sense of awe and bewilderment critical to an unquestioning faith in God. Even the vaunted Maimonides, in his depiction of Moses as a prophet who supposedly acquired all knowledge, had gotten it wrong, according to Hurwitz. There is no end to knowledge, for "in every generation new wisdom and understanding are revealed that were unknown to earlier scholars." Thus the designation "wise man" can only pertain to the knowledge accumulated in a particular era, and not in general. In the end, science demonstrates neither the mastery of nature by humans nor the veracity of their rational investigations, but quite the opposite. The more discoveries that science makes, the more mysterious the divine handiwork appears. The more we think we understand about the world, the more we appreciate how much we do not know. Scientific discoveries accordingly serve to make human beings more uncertain about their own capacities and more dependent on divine providence and the faith of their ancestors.[27]

THE SCHOLAR AND THE COMMUNITY

In one respect, Ulman and Hurwitz shared an important characteristic: a deep and passionate concern for the welfare of their community and an abiding sense of their capabilities as educators to lead their constituencies in the right intellectual and spiritual direction. But even so, the results of their efforts to influence their communities, or even to be noticed at all, are conspicuously different. Ulman died a lonely and embittered intellectual, feeling ignored and ineffectual in shaping a meaningful cultural agenda for the Jewish community. Hurwitz died a best-selling author, the aggressive marketing of his book having solidly paid off. Despite his modest origins, his encyclopedia became one of the most popular books among traditional Jewish readers, especially in eastern Europe. He connected to a readership broader and more numerous than Ulman no doubt ever dreamed of reaching. Thus, the ultimate impact their respective writings had, including their reflections on Jewish communal life and their self-image as Jewish educators, offers a final insight into this collective portrait of these two thinkers who almost met each other.

Naphtali Ulman's writings, especially those in manuscript, reveal an arrogant and sour temperament. His provocative justification for forgoing rabbinic approbations of his book is only one indicator of his condescending attitude. In various places in his writing, he complains about the poor rabbinic leadership of the community and its ignorance, which make Jews look stupid in the eyes of non-Jews. These leaders, he said, are "children with respect to their intelligence and grown-ups with respect to doing evil, always ready to harm good people among us to the extent they are able. They enjoy this role and they consider this a way of enhancing their worship of the almighty God."[28] In contrast, Ulman strived from his early youth, so he claimed, to do the right thing and to lead people in the proper direction based on sound principles of reason. He acknowledged the earlier contributions of sephardic scholars who wrote philosophy, but, he said, their work remained obscure and incomplete; Ulman therefore decided to finish their work. But his message was ignored or rebuffed, and the terrible leaders of his own time exiled such rational learning from the community so that the Jewish people became an embarrassment before the world. Publishing a book in Dutch as well, Ulman attempted to defend philosophical study among his people based on traditional texts and philosophical analysis. But his challenge to the communal leadership went unanswered; indeed, no one even acknowledged his singular contribution.[29]

The major targets of Ulman's stinging barbs were those Jews of his generation who presented themselves as pietists (*Mitḥasdim*), obscuring the teachings of the Torah in their riddles and metaphors and hinting at hidden secrets that never existed. Their emotional piety had replaced serious rational inquiry, and they were dangerous to Judaism, since piety without honest investigation is not piety at all, claimed Ulman. His most biting and sarcastic observations of these Jewish leaders is found in a small composition called *Ma'aseh ha-Tartuffe*, whose title was obviously borrowed from Molière's play.[30]

The manuscript, copied in 1776 in the Hague by someone who had heard an oral presentation by Ulman, offers a large inventory of alleged hypocrites among the Jewish people. They include men who make spectacles of themselves—wearing larger prayer shawls, praying with their eyes closed, and raising their voices in prayer so that they might be noticed. After prayer, they read from a kabbalistic book, making scary sounds and motions while recounting ridiculous stories and dream visions. They reveal excessive concern for halakhic minutiae while ignoring the serious Torah scholar. They are arrogant, foul-mouthed, and run after money, food, and drink. They honor only the rich and ignore the poor. Their hide their ignorance by responding to their critics with aggressiveness and anger, just as Catholic priests act in vilifying Jews. Ulman may have been referring to the Ḥasidim of eastern Europe in these biting remarks but probably had in mind a larger group, including Jews closer to his own environment, especially leaders of the community who were illiterate, immoral, and displayed their bad qualities in destroying the social fabric of Jewish society.

By composing such a diatribe, Ulman openly displayed his anger and frustration with the intellectual and social morass into which he imagined the Jewish community had fallen. But there was more to his critique than mere vituperation. He also offered a way out, if only the community would embrace him and philosophers like him, community-minded and morally conscious men who cared genuinely about the fate of their people. What leaders required at this moment of crisis, Ulman contended, were intellectuals armed with a knowledge of philosophy and history, especially the wisdom of the medieval philosophers. Only such men could face up to the intellectual challenges of their day both within the Jewish community and beyond it. Ulman saw himself as a champion of the Jews much like the tenth-century Saadia, who successfully combated the Karaites, the enemies of rabbinic Judaism. Philosophy, he finally pleaded, was not the enemy of the Jewish community; immoral behavior and bad character were. The sermon closes with the following: "These are the words of the bitter and sad . . . Naphtali called Herz Ulman of Mainz."

Pinḥas Hurwitz also had enemies, and he was not reluctant to castigate them in print, beginning with the publisher who had released a pirated edition of his book. He also attacked deists and atheists such as Voltaire and Spinoza, and he was particularly enraged by two Maskilim, Saul Berlin and Isaac Satanov, who masked their heretical views in traditional-sounding rabbinic writing, thereby misleading and confusing a naive community of readers.[31] Hurwitz never openly attacked Ḥasidim but only the followers of Shabbetai Ẓevi and that group's leaders Abraham Cardoso, Nehemiah Ḥayon, and Berakhiah Russo, who, Hurwitz charged, undermined the very foundations of the Jewish faith and sullied the good name of the Jewish community among its neighbors. Furthermore, the Sabbateans had given a bad odor to the kabbalah itself, confusing a pure and sacred lore with their heretical notions.[32]

Hurwitz, like Ulman, also offered his own social criticism of certain practices that he observed in the community. He decried those Jews who refused to teach their children an honest living and expected them to study the Torah alone. He was offended by the class of lazy itinerant and penniless scholars who refused to involve themselves in productive labor. He also criticized the rich for their ostentation and for ignoring their social responsibilities. He offered a catalogue of moral excesses of his day, including jealousy, lust, frivolous conversation, and pride, but especially singling out the senseless hatred between Ashkenazim and Sephardim.[33]

The apparent similarities between Ulman and Hurwitz regarding their visions of Jewish life are obvious. Both displayed a strong impulse to address social ills and to provide moral leadership to their communities. They were also alike in their mission to inculcate a new intellectual agenda through their books and teaching—in the case of Ulman, a new metaphysics, and in the case of Hurwitz, a fresh accounting of natural philosophy and scientific discoveries. In the end, however, they remained worlds apart. Ulman departed the earth "a bitter and sad" man, frustrated by the gap between his high expectations of himself and his inability to impact the community he wished to educate. Ultimately, he caused his own marginalization and alienation from the community by lashing out at his supposed enemies through bitter sarcasm and castigation of its religious and political leaders.

In contrast, Hurwitz constructed a message that was uplifting, personal, and spiritual. The style of *Sefer ha-Brit* was engaging and charming. He invited his readers to experience a journey through the wonders of nature as well as an ascent to the mysteries of the divine spirit itself. Along the way, he addressed his audience directly, encouraging them to read his book in the proper order and patiently leading them from point to point, from subject to subject. In so doing, he succeeded in imbuing the encounter with the outside world with a mystical and moral aura. Natural philosophy was wrapped in the garments of Torah and all led to the ultimate revelation of the divine, which was the culmination of prophecy itself. Ulman hated Jews like the Ḥasidim who made a spectacle of their piety, while Hurwitz never mentioned them explicitly. And ironically, his book was received with great passion by Ḥasidim and Mitnagdim alike. It became a kind of holy book embedded within the traditionalist camp rather than outside it.

In the final analysis, it was not only Ulman's metaphysical assumptions that were out of style with a new age; it was also his rage, his impatience with displays of piety, and his indifference to the spiritual needs of those he wished to serve that left his intellectual project almost totally forgotten in the end. Ulman's works, unread and unpublished, provide clear testimony to his failure as an educator and writer. Hurwitz, in contrast, the consummate salesman and entrepreneur, understood his role well: to entertain and to inspire while providing a powerful bridge between natural philosophy and *musar* (ethics), between intellectual edification and moral and spiritual improvement. The combination proved to be his greatest success. Let us now turn to his life and best-selling book.

CHAPTER 2

Pinḥas Elijah ben Meir Hurwitz

TOWARD A BIOGRAPHY OF A POPULAR AUTHOR AND AGGRESSIVE BOOK DEALER

HAVING introduced Pinḥas Hurwitz by way of a virtual dialogue with Naphtali Ulman in chapter 1, it is now worth considering more extensively his biography and popular work. Reconstructing his peripatetic life, to the extent that that can be done, and analyzing the process by which he wrote, printed, and distributed *Sefer ha-Brit* are the goals of this chapter. But such tasks are not simple. While Hurwitz offers a great deal of autobiographical information in his book, it extends only over a limited period of some ten to fifteen years. During his early years and after the initial publication of his famous book, he becomes virtually invisible. Moreover, even in the period of his life that he does document, significant gaps make it difficult to determine the precise itinerary of his travels from east to west and back and the people he contacted throughout his wanderings.

Even the date and place of Hurwitz's birth cannot be established with any degree of certainty. He appears to have been born in Vilna to Meir and Yenta Hurwitz, themselves residents of the city, and the year of his birth is usually given as 1765, but no strong evidence for either place or year is forthcoming. Noah Rosenblum refers to Hurwitz's mention of two lunar eclipses, one in The Hague and one "in Vilna, in the state of Lithuania, the city of my birth." He also witnessed a solar eclipse in 1769, the earliest life event he records. If he were a young person who had actually observed and remembered this astral occurrence, he might have even been born earlier than 1765. But all this is mere conjecture. What seems certain is his matter-of-fact statement that Vilna was the city of his birth.[1]

Whether Hurwitz was born in Vilna or not is of great significance for understanding his cultural origins. While most historians take him at his word that he hailed from Vilna, ambiguity remains: was he referring to his parents' origins or to his own? Nathan Gelber, the historian of the Jews of Buczacz, offers this scenario: "During these years, there lived in Buczacz R' Pinkhas Eliyahu Horovits author of *Sefer HaBrit*. His father Meir and his mother Yente were from Vilna. In 1776 [clearly the wrong date!] their son was born in Lvov, through which they were passing on their way to Buczacz, where they wished to settle. However, they called him 'Vilna' after his father's descent."[2] Gelber based his claim on the testimony of Gershom Bader, who reported that he had heard from elders in Cracow who knew Hurwitz personally that he was born as his parents passed through Lemberg (Lvov) to Buczacz, but he nevertheless followed his parents' lineage from Vilna.[3]

If indeed Hurwitz was not born in Vilna, this might explain why he goes largely unmentioned in historical discussions of the cultural life of the famous Jewish city during the late eighteenth and early nineteenth centuries. Virtually the only historian to place Hurwitz in the history of Vilna Jewry was Samuel Joseph Fuenn (1818–1890). He devoted a long entry in his history of the Jewish notables of the city, *Kiryah Ne'emanah* (Faithful City, 1860), to the "famous rabbi" Pinḥas Hurwitz, referring to the common confusion of identifying the author either as the Gaon of Vilna or as Moses Mendelssohn, a confusion Hurwitz himself mentioned in the second introduction to the second edition of his book. Perhaps such a characterization echoed the well-balanced mixture of enlightenment and traditionalism that Fuenn found attractive in Hurwitz's book. He closed his entry on Hurwitz by claiming him a worthy member of the Jewish community of Vilna.[4]

If Fuenn's claim regarding Hurwitz was correct, it does seem odd that virtually no other evidence besides Hurwitz's casual remark exists to link him to this thriving Jewish metropolis, especially during the lifetime of its most illustrious rabbi, Elijah the Gaon of Vilna.[5] Of course, Hurwitz himself proposed a linkage to the Gaon by remarking that some thought that actually Elijah was the author of *Sefer ha-Brit*.[6] Was this simply a gratuitously self-promotional comment by Hurwitz, intended to make the book seem more important in the eyes of traditional Jewish readers, and therefore more saleable? This may well be the case, although some grounds do exist for associating Hurwitz's book with the Gaon and his circle, whether Hurwitz was actually born in Vilna or not. In the first place, there is a telling connection to Barukh Schick, the author of a Hebrew translation of Euclid, prepared explicitly with the blessing of Elijah Gaon himself. This translation, published in The Hague in 1780 with the *haskamah* of the ashkenazic rabbi of Amsterdam, Saul Loewenstamm, himself a relative of the Gaon, indicates Loewenstamm's strong interest in the sciences. That this same rabbi wrote a similar *haskamah* for Hurwitz's work some nine years later appears to be more than coincidence. At the very least, it suggests that a close associate and relative of the Gaon promoted two Hebrew works on the sciences written within a ten-year period, with the probable approval of the eminent Vilna rabbi as well. That Schick chose The Hague to publish his book is also noteworthy, given Hurwitz's own intimate relation to the city, where he both completed much of the writing of *Sefer ha-Brit* and secured the rabbinic approbations he needed to successfully market his book.[7]

Hurwitz's relationship to Schick was more immediate than that, however. Not only did he cite Schick's *Ammudei ha-Shamayim ve-Tifferet Adam* (Pillars of Heaven and the Splendor of Man, 1776–77) in *Sefer ha-Brit,* but he also repeatedly copied from it.[8] Schick's science book was only one of the many Hebrew sources Hurwitz "borrowed" from, either openly or not, but he certainly was influenced by the work and might even have met the author himself in Berlin, a city both men visited around the same time.

Aside from Schick, although Hurwitz's intellectual connections to the Gaon's close associates and students are almost never explicit, nevertheless, they might be plausibly suggested. Historians have long debated the degree of the Gaon's commitment to the study of the sciences, not to mention philosophy. There seems no doubt, however, that some of his most prominent students were fascinated by natural history and drew from the sciences in their own Hebrew writing. Most prominent in this regard was Abraham, the Gaon's own son, whose composition *Gevulot ha-Areẓ* (The Boundaries of the Land, 1801) was essentially the partial translation of a popular work by the naturalist Georges-Louis Leclerc, comte de Buffon (1707–88).[9]

Other students of the Gaon with a particular interest in the sciences included Benjamin Zalmin Rivlin (1728–1812) and Isaac Ḥaver Wildman (1789–1853), disciple of the Gaon's close associate Menahem Mendel of Shklov (1750–1827).[10] In the case of Wildman, it is certain that he actually read *Sefer ha-Brit,* for he cited it in his *Magen ve-Ẓinnah* (The Shield and the Target), referring specifically to Hurwitz's long chapter titled "The Way of

Faith." At the beginning of the same chapter where this citation is located, he also mentions Kant's claim of the uncertainty of metaphysical knowledge in contrast to the certainty of kabbalistic sapience. Alan Brill has suggested Julius Barash's *Oẓar ha-Ḥokhmah* (The Treasury of Wisdom) as the source of Wildman's understanding of Kant, but Hurwitz's extensive discussion of Kant's challenge to philosophical metaphysics (discussed in chapter 4) seems to fit more precisely as the source of his discussion.[11] Indeed, this attribution seems almost certain given that Wildman cites the very chapter in *Sefer ha-Brit* in which Hurwitz discusses Kant.[12] Assuming Hurwitz was the primary source of Wildman, this would also betray his likely influence on other nineteenth-century Jewish thinkers such as Ẓadok ha-Cohen of Lublin (1823–1900) and Israel Salanter (1810–83), both likewise mentioned by Brill.[13]

If Hurwitz's Vilna pedigree remains blurry while his intellectual connections to the city appear more apparent, his subsequent childhood and young adulthood are totally obscure. It is impossible to ascertain where he lived before his departure for western Europe around 1789. One likely possibility is that he took up residence in Buczacz but eventually moved to Lemberg for medical reasons. In the first introduction to *Sefer ha-Brit*, he reports that because of his exhaustive work on the book, he severely strained his eyes in Buczacz and was obliged to leave to receive treatment in Lemberg:

> At the beginning of my writing this book I was in Buczacz in the region of Galicia. Out of soulful desire and wonderful urge to compose this work, I worked quickly and incessantly writing the book day and night in darkness and in light with great diligence and mostly without sleep for half the night. Because of the writing, my eyes dimmed, the light from the pupils of my eyes was removed so that I couldn't see anything. I collapsed in bed while my head and eyes were very impaired because of my sins. I took medications there throughout the winter, but no one in the congregation of Buczacz could offer me any cure. So I traveled from there to Lemberg, which was the capital of this region, and I took medications there for a half year in the wonderful rabbinic home of the noble Naḥman Reise. Since I hoped for light and there was none, and I was almost despairing of ever seeing again, I remembered the words of the saying, One should make a vow in a time of suffering, so I vowed to the God of Heaven that if God is with me and he restores my eyesight and heals me so I can resume writing my book and God will restore it to be completed, I will not attach my name to it explicitly as is known to the people of Lemberg. And blessed is God my

> light and redeemer who heard the voice of my petition, who sent his word to heal me so that God opened my eyes. His mercy encompassed me by giving me good health and excellent sight. I thank you the Lord my God with all my heart and I will honor your name forever, since your kindness is great to me so that after this episode I was able to write some of my book there [in Lemberg] and afterwards I was able to write another part in The Hague, the capital city of the state of Holland as it is explained in the [rabbinical] approbation from there.[14]

Based on these remarks, it appears that Buczacz was a sort of launching point for the writing of Hurwitz's book, and when he overcame his medical problems in Lemberg, he continued to write there as well, although he acknowledges that he did not finish but continued to write when he arrived later in The Hague, from where he concluded a part but not the entire composition. In Lemberg he was welcomed into the home of an apparently well-to-do Jew named Naḥman Reise (or Reiss). In the first edition of *Sefer ha-Brit*, one finds the addition that this Jew was at the court (*ḥaẓer*) of Itsik Matil (or Mottels).[15] Whatever the specific identities of Reise and Matil, it is already clear at this early stage of Hurwitz's career that he sought and obtained the economic and social support of well-to-do Jews wherever he traveled. His ability to complete so challenging a book rested on the security and hospitality these individuals provided him. When the book was eventually published, such individuals also provided crucial assistance in buying the book and selling it to others.

Hurwitz concludes his discussion of the recovery of his eyes by mentioning that the rabbi of Oven wished to release him from his vow, cautioning him that it might harm him when it came to selling the book, since someone might falsely claim that he was not the author. R. Moses Münz of Brody (ca. 1750–1831), rabbi of Oven (Ofen = Buda or Obuda), was probably the counselor to whom Hurwitz referred. Elected rabbi of Obuda in 1789, he gave approbations to other important books, including the 1818 edition of the *Biur* of Moses Mendelssohn and the Hungarian reform rabbi Aaron Chorin's *Emek ha-Shaveh* (Valley of the Plain) of 1803. He not only was one of the seven rabbis who wrote *haskamot* for *Sefer ha-Brit* but was also one of its primary distributors, as we shall see below. Whether Hurwitz ever visited him in Obuda is unclear.[16] In the end, Hurwitz did not follow the rabbi's advice, but honored his vow and omitted explicit mention of his name in the first edition (though it included obvious hints). In any case, he no doubt assumed that many readers would recognize his identity, including the rabbis who allowed their approbations to be published in this otherwise "anonymous" work.

TO THE WEST: BERLIN AND BEYOND

It appears that during Hurwitz's Galician sojourn he visited, at least once and perhaps on several occasions, the city of Berlin. Rosenblum remarks on Hurwitz's strong connections to Prussia—his heavy usage of the German language in describing scientific terms and his intellectual reliance on prominent figures of the German *Haskalah,* such as the aforementioned Barukh Schick and Barukh Lindau, both of whom Hurwitz utilized by openly citing them or simply copying them without acknowledgment. Hurwitz knew that Lindau, in writing his own scientific work *Reshit Limmudim* (The Beginning of Studies) in 1788, had primarily copied the book from George Christian Raff, so "borrowing" from Lindau was essentially "borrowing" from Raff. He also was familiar with the Hebrew work of Solomon Maimon; and he excoriated the writings of two other Maskilim, Saul Berlin and Isaac Satanov, who were certainly connected with the city of Berlin.[17]

Hurwitz's primary contact in Berlin was Ẓevi Hirsch Levin (1721–1800). In 1758, Levin was appointed rabbi of the Great and Hambro synagogues, London (where he was known as Hart Lyon), holding the post for seven years. In 1764, he relinquished his position and went to Halberstadt, Germany, where he was appointed rabbi and *rosh yeshivah.* In 1770, he was appointed rabbi of Mannheim, and in 1773 he became rabbi of Berlin. Levin was generally open to the *Haskalah* and initially developed a strong friendship with Moses Mendelssohn. They parted company, however, over their respective reactions to the Hebrew manifesto of the *Haskalah* of Naftali Wessely, published in 1782. Despite his more conservative position, Levin supported his son, Saul Berlin, over the allegations that he had forged the rabbinic work *Besamim Rosh* (Berlin, 1793).[18] This did not seem to impair Hurwitz's appreciation of the father, notwithstanding his strong reservations about the son.[19]

Ẓevi Hirsch Levin's brother was Saul Loewenstamm, the ashkenazic rabbi of Amsterdam, who was the first to formally endorse Hurwitz's book based on the warm recommendation he had received from Levin. In the words of Loewenstamm, Hurwitz "showed me a holy epistle from my brother the great eagle, jewel of the age, Rabbi Ẓevi Hirsch . . . the head of the rabbinic court of the holy congregation of Berlin . . . who wonderfully praised him for his mastery of the secrets of the true kabbalah and who had acquired much wisdom to understand and to teach something learned of their subjects."[20]

That Hurwitz frequented the home of the chief rabbi of Berlin before his departure for the Netherlands is attested by an unusual Hebrew source, one of the earliest references to *Sefer ha-Brit*. Moses ben Eliezer Phoebus Koerner (1766–1836) was rabbi in Rendsburg (Schleswig), Shklov, and Grodno, and, very much like Hurwitz, he traveled extensively in his later years to promote his own primarily homiletical compositions. He eventually settled in Breslau, where he died. Koerner obviously knew Hurwitz personally, as he relates in the opening page of his *Ke'Or Nogah* (As the Radiant Sunlight [after Proverbs 4:18], 1816), composing, so he writes, "a letter of hidden love to the author of *Sefer ha-Brit*." He refers to Hurwitz as "an exalted friend, my close friend, the perfect wise man, investigator and fearer of God" and mentions that they had met sometime during the period he published his book *Zerah Kodesh* (Holy Seed, 1798) at the home of Rabbi Levin. If Koerner actually meant that the two had met in 1798 in Levin's house, the date falls considerably later than the period prior to Hurwitz's departure for the Netherlands—some nine years later, to be exact. But Koerner was recalling a meeting or more likely a series of meetings some twenty years earlier and was probably not precise. Whatever the time frame of their meeting, Koerner confirms Hurwitz's strong ties to the Berlin rabbi and allows us to speculate plausibly that Hurwitz met not only Koerner there but also other significant guests, such as Lindau, Schick, and other like-minded Jewish intellectuals. It suggests Hurwitz's extraordinary ability to attract wealthy and powerful rabbinic and lay leaders to him and his intellectual project, a major ingredient of his great success in publishing and selling his book.[21]

By 1789, Hurwitz had departed either Galicia or Berlin for Holland. He apparently prepared his trip carefully, bearing the letter from Rabbi Levin to present to his brother-in-law Saul Loewenstamm in Amsterdam and thus initiate further introductions to the other rabbis of Holland. As we saw in the previous chapter, Hurwitz, over the course of three years, secured seven *haskamot*, four from rabbis in Amsterdam, Rotterdam, and The Hague, and three from rabbis in Cracow, Obuda, and Pinczow. His success in obtaining all seven rested on the already existing network the rabbis had with one another. Until his death in 1790, Saul Loewenstamm was the central figure in the group. David Azevado was his junior colleague in Amsterdam. Judah Leib Mezerich in The Hague had been ordained by Loewenstamm, while Isaac ha-Levi and Aryeh Leib Breslau were both from the same family that included the well-known ashkenazic rabbi of Amsterdam Ḥakham Ẓevi and his even more famous son Jacob Emden of Altona. The other two, Moses Münz and Isaac Abraham of Pinczow, were either familiar with Hurwitz directly or had heard of him and his project through one of the other rabbis. In finally choosing to settle in The Hague for at least an entire year, as Rabbi Mezerich relates in his approbation, Hurwitz had prepared himself well in establishing the proper conditions for writing his book. In The Hague, he finished at least the first part if not more, which Mezerich had seen with his own eyes.[22]

Hurwitz made various references to The Hague and to Holland throughout *Sefer ha-Brit.* He speaks particularly about the climate, the vaporous air, and the frequent fogs. He marvels at the dikes built to contain the flooding of the area, and he describes various animals and fish and even a solar eclipse that he observed with his own eyes.[23] Hurwitz also possibly commented on the social dynamics of Dutch Jewish society, particularly the tense relations between Sephardim and Ashkenazim in Amsterdam itself:

> The Sephardim hate the people from Germany and Poland with a great hatred, despite the fact that they live in the same city with Ashkenazim. They praise their own families who are descendants of [the tribe of] Judah and [ask], "What do we have to do with the inferior and despised Tedeschi?" While, in direct opposition, the German and Polish Jews refer to them [the Sephardim] as bitter families in their appearance and in their evil and nasty actions, while "we, the sons of Ephraim, come from an exalted family, from the tribe of Ephraim." Thus this senseless hatred is more difficult than that between one person and his family member or his fellow residents in the same land.[24]

Hurwitz's observation about the tense relations between Sephardim and Ashkenazim is particularly relevant to his larger discourse on loving neighbors (from which this passage is taken) in the second part of *Sefer ha-Brit,* which we shall examine more thoroughly in a later chapter. The theme of the entire chapter is universal love of all human beings, but Hurwitz ultimately shifts to describe "senseless hatred" among the ethnic groups within the Jewish community itself and there explicitly mentions the tensions of these two subcommunities. It would appear that Hurwitz's exposure to the intracommunal tensions in the mixed Jewish communities of Holland provided food for thought, leading him to pen an entire treatise on the need for human love as the penultimate chapter of his book.

Rosenblum cogently asks why Hurwitz chose to settle in The Hague rather than Amsterdam given the latter community's larger size and its higher concentration of Ashkenazim. He proposes most plausibly a possible connection with the wealthy family of Tobias Boas (1692–1782) and especially the son, Abraham, whom Hurwitz mentions explicitly in *Sefer ha-Brit,* in reference to a letter written by the Jews of Kuchin requesting religious books, a letter that Hurwitz himself had seen.[25] This reference might suggest that Hurwitz visited this wealthy household and may have been supported by the Boas family while he completed a large part of his book in The Hague. Yet in the many references to the Boas family in the communal ledger of The Hague, Hurwitz is not mentioned.[26] Still, given the hospitality the Boas family extended to other learned visitors to their city, such as Ḥayyim David Azulai, who passed through The Hague in 1778, such a supportive connection seems possible, even despite the fact that Hurwitz was an ashkenazic Jew and the Boas family was sephardic. Certainly Boas's intellectual interests in the sciences and in new discoveries would have made Hurwitz an attractive houseguest for him.[27] A less likely reason for Hurwitz's attraction to The Hague is that he wanted to meet the aforementioned Naphtali Ulman, who died suddenly before Hurwitz's arrival, or that he lived and was supported by the rabbi Judah Leib Mezerich, who wrote one of the endorsements of his book.

Based on the *haskamot* of the four Dutch rabbis, it seems probable that Hurwitz lived in The Hague for at least a year in 1789–90. He may have returned to Poland a year or two later to gain the last of his endorsements from two Polish rabbis. In the interim, he may have journeyed to London, where he remained for an unspecified amount of time. Hurwitz mentions explicitly a journey on the sea, most likely from Amsterdam to the coast of England. He also describes examining a book in London, shown to him by a sephardic scholar.[28] More significant is the close relationship he describes in *Sefer ha-Brit* with one of the community's leaders at the end of the eighteenth century, Eliakim ben Abraham, known as Jacob Hart (1756–1814). This appears to be more than a passing relationship; Hurwitz refers to Hart as one "who loved my soul, my friend, a delightful man, an exceptionally wise man outstanding in Torah, the late noble rabbi Eliakim Gotshlick Hart ben Abraham of London."[29]

Hart, a jeweler and Hebrew writer on religion and science, was born in London and, except for a four-year sojourn in Germany, where he may have attained his rabbinic ordination and perhaps met Hurwitz for the first time, spent his entire life in his native city.[30] He earned a respectable living working on the Strand as a jeweler, which allowed him to generously support the synagogue he helped to establish, the Denmark Court Synagogue. One indicator of his social and financial position was that at his death, three coaches were sent to the synagogue for his funeral at double the usual cost. Hart was also a member of Hiram's Lodge, Swan Street, home of the modern Freemasons. In 1801, he is listed in the Lodge's directory along with his age (forty-five) and profession (silversmith).

Beyond his professional career, Hart aspired to write Hebrew books on a wide variety of subjects, including grammar, mathematics, natural philosophy, and kabbalah. His major intellectual project was a planned series of ten relatively modest works entitled *Asarah Ma'amarot* (Ten Essays), of which only five were printed. The first, *Milḥamot Adonai* (Wars of the Lord), on the subject of Newtonianism, was published in London in 1794 and is his most important work. A year later he published *Binah la-Itim* (Chronology), an explication of the Daniel prophecies in light of past chronologies and contemporary political events. His other writings include a summary and commentary on works of Joseph Delmedigo, a seventeenth-century Hebrew writer on kabbalah and science; another summary of kabbalah; a work on the Hebrew vowels; and a Hebrew grammar, the latter three published in Germany. His other apparently unpublished works dealt primarily with kabbalah, especially the system of the sixteenth- century kabbalist Isaac Luria, and with mystical speculations concerning the secrets of numbers. Hart's entire opus explores the cultural links between natural philosophy and mystical theosophy, an integration found also in the writings of the aforementioned Delmedigo, Newton and some of his disciples, and obviously *Sefer ha-Brit*.

The *Milḥamot Adonai* is ostensibly a polemic against four groups who corrupt the beliefs of traditional Jews and impugn the sanctity of their sacred revelation: ancient pagan chroniclers who claim that other cultures preceded that of the Hebrews, Aristotle and his followers, Descartes and his followers, and Newton as interpreted by his more radical students. The most sustained part of the polemic is his treatment of Newton, whom he initially praises both for his scientific contributions and his commitment to the truth of biblical prophecy; but ultimately he criticizes the Latitudinarian and deist followers of Newton for their assault on divine providence and revelation. Hart explicitly mentions Newton's commentary on the Book of Daniel, which suggests that it was probably the model for Hart's own commentary as well.

Hurwitz cites Hart four times in *Sefer ha-Brit*. In the first instance, in the same passage in which he singles him out as a loving friend, Hurwitz describes how Hart solved the difficult challenge of interpreting the particular wording of the blessing of the new moon when referring to the heavenly firmaments. In the second instance, he quotes a letter from Hart written after the first edition of *Sefer ha-Brit* had appeared praising the book but objecting to Hurwitz's questioning of the Copernican position, since the earth's movement was so insignificant in relation to the universe as a whole. The third time he cites Hart is to offer the latter's interpretation of a Zoharic passage with which Hurwitz wholeheartedly agrees. Finally, he mentions that he consulted unsuccessfully Samuel Auguste Tissot's popular history of medicine in the library of his friend Hart for an allusion to smallpox.[31] These scattered references reveal a deep friendship based on an appreciation both of the sciences and of kabbalah, and particularly the ability to freely integrate the one with the other. Both men loved science but were fearful that its radical interpreters might undermine divine providence and creation. In the same way that Hart lauded Hurwitz's book overall yet declared Hurwitz's ambivalence to Copernicus silly and unnecessary, he likewise praised Newton while at the same time distancing himself from too mechanistic an interpretation of his theories, which might diminish God's role in creation. It seems likely that Hurwitz journeyed to London to be with his friend and that Hart received him with warm hospitality and generous financial support. Hurwitz had again found a rich patron and intellectual partner sympathetic to his project and his personal needs.

How long Hurwitz lived in London remains a mystery, although it is reasonable to assume that the situation Hart offered him was quite satisfactory. As has been mentioned, it is not clear when Hurwitz left The Hague or if he returned there after his London visit. Noah Rosenblum believes he then left for eastern Europe, specifically Cracow and Pinczow, based on the dates on which the two rabbis from these places signed their approbations for his book, that is, 1792 and 1793.[32] This is certainly plausible but not sure. Given the networks of communication among the seven rabbis, Hurwitz's personal appearance in each city may not have been required.

PRESSBURG SOJOURN

Rosenblum is on more solid ground, however, in placing Hurwitz in Pressburg, Hungary, given this remark from the first edition of *Sefer ha-Brit:* "I wrote a part of this book in the holy congregation of Pressburg in 1794."[33] Hurwitz's primary reason for going to Pressburg was again a personal acquaintance, this time Beer Ben Isaac Oppenheim (1760–1849), rabbi and Maskil, letter writer and author of a responsa collection entitled *Mei Be'er* (Waters of the Well, 1829).[34] Oppenheim showed him the manuscript of this composition and asked him to publish a small section of it in his forthcoming volume, to which he agreed, commenting: "And I saw a lovely book filled with wisdom, a manuscript of a wonderfully wise rabbinic scholar, a precious man with great understanding, and a writer, he is the honorable rabbi Beer Oppenheim, a resident of this aforementioned city [Pressburg]. He requested that I publish in my book a small treatise from his [manuscript], and I promised that I would and here it is as it was written in its original language."[35]

Oppenheim's homily on a rabbinic passage, which extols a conventional inductive empiricism over deductive a priori reasoning, is hardly as brilliant as Oppenheim might have thought. As Rosenblum points out, Hurwitz probably agreed to assume the expense of publishing these extra pages as a gesture of gratitude to his host, who offered him support and intellectual companionship during his sojourn in Pressburg.[36] In the end, little in Oppenheim's passage indicates a profound knowledge of either rabbinic literature or philosophy.

This makes all the more ludicrous the suggestion of Isaac Hirsch Weiss, whose wife was an Oppenheim, not only that Hurwitz lived in Oppenheim's house but also that Oppenheim had translated all the secular works he consulted in writing *Sefer ha-Brit*.[37] Although Oppenheim's homily certainly fits in with the spirit of Hurwitz's work, it offers little intellectual benefit or novelty. But Oppenheim's financial backing was no doubt an advantage to Hurwitz, as were his rabbinic connections. It was probably because of his friendship with Oppenheim that Hurwitz secured the last of his rabbinic approbations, this one, as we have seen, from Moses Münz of Brody, rabbi of Oven, in 1795.

Rosenblum was thus quite accurate in his identification of the Oppenheim passage in *Sefer ha-Brit*, in describing the relations between Hurwitz and his Pressburg benefactor, and in discounting Weiss's exaggerations about Oppenheim's alleged influence on him. He erred, however, in underestimating the impact Hurwitz had on the more important rabbi of Pressburg in later years, Moses (or Ḥatam) Sofer. Sofer, of course, was not yet in Pressburg in 1794, nor was he on good terms with either Oppenheim or

Münz. But he did come to appreciate profoundly *Sefer ha-Brit* later on, as will be discussed in chapter 6, and he even encouraged its author to change a passage in the book that he found religiously objectionable.[38]

After Pressburg, Hurwitz's whereabouts are highly conjectural. Did he travel to Buda to gain the *haskamah* of Rabbi Münz in 1795? Did he head for Brünn to oversee the publication of his book in that same year? Given his deep concern over how the book was to be printed, as Rosenblum reminds us, wouldn't it have been logical for him to invest considerable time and energy in that process?[39] This indeed may be the case, but it is also obvious that Hurwitz became obsessed with the fine points of printing primarily after losing control of his publication with the appearance of a later pirated edition. It is primarily, if not exclusively, after this harrowing experience that he involved himself in every stage of the book's printing. Thus, whether he was actually in Brünn at all before 1797 remains uncertain (he does seem to have been there later on, as we shall see). Unfortunately, a precise accounting of his whereabouts from the mid-1790s until his death in Cracow in 1821 cannot be reconstructed. He clearly traveled extensively, selling his book. At some unknown date, he seems to have settled finally in Cracow, where he eventually died and was buried and where one of his books was published long after his death.

Ironically, at the point when Hurwitz's biography becomes increasingly inaccessible, another narrative becomes most visible, namely the history of his book. Hurwitz had little to say about himself in the wake of *Sefer ha-Brit*'s first publication in 1797, but he had much to relate about the process by which he produced the book, attempted vainly to protect his rights as an author, and ultimately reinvented his book to become a best-seller. It is to this story that we now turn.

PUBLISHING AND SELLING *SEFER HA-BRIT*

Hurwitz finally succeeded in publishing his book in Brünn in 1797 at the printing house of Joseph Karl Neumanns and Joseph Rossmann, about whom we know nothing beyond their names. Why Hurwitz published the book in Brünn, a seedbed for heretical groups, especially the Frankists, also remains a mystery. As we have seen, he did not disclose his authorship, although several hints in the text reveal his identity to the discerning reader.

The work was published in two parts: the first primarily but not exclusively focuses on natural philosophy, while the second deals with spiritual and moral issues.[40] Rather confusingly, the book starts off with two pages enthusiastically approving the work, in German and in Yiddish, by the well-known Prague censor Karl Fischer, dated January 21, 1799. Were these pages appended after the book had circulated for almost two years, without having received the permission of a censor? The seven *haskamot* of the rabbis Hurwitz solicited follow. All of them strongly endorse the book and offer stern warnings not to publish the work without the permission of the author. The second part closes with a list of errata, followed by Beer Oppenheim's homily, which was removed from all subsequent editions.

Karl Fischer maintained cordial and professional relations with the rabbis of Prague and was well known as a promoter of Hebrew books. His role in the dissemination of Jewish culture in Prague stemmed from a deep commitment to humanity as a whole. As he wrote in a Hebrew letter to Rabbi Eliezer Fleckeles in 1812, "I said: 'Anyone who speaks the truth, likes justice, and follows the path of the sincere, whether he be Jew, Christian, Greek or Muslim, is eminent and worthy of love [*hu ḥashuv ve-ra'ui le-ahavah*].'" He undoubtedly appreciated Hurwitz's similar sentiment as expressed in his long section on loving humanity.[41]

It is unclear how Hurwitz went about securing Fischer's glowing endorsement. Whether he actually met the censor before the publication of the book is unknown, but less than two months after Fischer wrote his approbation, on March 4, 1799, Hurwitz addressed a letter to him. The letter is fascinating for several reasons but most significantly because it was penned in German, a language Hurwitz had claimed he never learned, thus obliging him to work with a translator when gathering materials for his book. The letter found in the Karl Fischer archives in Prague may have been a German translation of a Hebrew original (though Fischer both read and wrote Hebrew), perhaps translated on Hurwitz's behalf by an associate, though it is possible that Hurwitz wrote it in German in the first place.[42]

Hurwitz opens the letter by thanking Fischer for his recent authorization of the publication of *Sefer ha-Brit* and promising him a personal copy of the book, to be delivered to him by an associate, a Polish Jew. He mentions that he had previously spoken to Fischer in person, indicating that he had visited Prague. He writes now from Brünn, he says, explaining that he crossed the border into Moravia, leaving five copies of his book with a good friend from Fuerth to be sold on commission. The friend had sold the lot in Pilsen (ninety kilometers west of Prague) and in Prague for a relatively cheap price. Hurwitz states that his agent in Prague has informed him that all copies had been sold there, and he needs to ask his publisher in Brünn to print more copies. He therefore wonders if Fischer might inquire about the availability of the book in the Prague area so that he can properly supply more books as needed. Despite the humble manner in which Hurwitz addresses Fischer, it seems presumptuous to ask the censor of his book to assist him in marketing it—but that appears to be precisely what Hurwitz was doing.[43]

The letter reveals a network of agents working for Hurwitz to sell his book—a Polish Jew, a friend from Fuerth, and even potentially the vaunted Hebrew censor of Prague! This is substantiated by a notice at the end of the first edition of the book, printed on the very last page: "This book is made available for purchase in the city of Brünn through R. Asher Garkach[?]; in Vienna through the noble R. David Leib Fered; in Pressburg through R. Beer Oppenheim; in Oven through the head of the rabbinical court; in Cracow through the author; in Prague through the noble R. Zelig Meliẓ; in Breslavia through the sons of the late R. Michael Mas; in Lemberg through R. Naḥman Reise; in Lublin through R. Fischel and R. Ziskindish; and in Nicolsburg through the noble rabbi Beer Herzilish."[44]

Most of Hurwitz's agents remain unknown to me, though the names Naḥman Reise and Beer Oppenheim are certainly quite familiar. The others were no doubt close associates, including some rabbis, and all were willing to facilitate the selling of Hurwitz's book. At the time the notice was written, Hurwitz was in Cracow—which might solve the puzzle of his whereabouts around 1797. Perhaps he had taken up residence in Cracow and only left when marketing his book. Taking this notice together with the letter to Karl Fischer, one gets the impression that from the beginning, Hurwitz did everything in his power to sell his book, working with a wide range of agents from Prague to Cracow, Lublin to Brünn, Nicolsburg to Breslavia.

To this evidence we might add an interesting remark presented in a lengthy review of *Sefer ha-Brit* in the Hebrew journal *Ha-Me'asef*, published in 1809 but written much earlier, soon after the first edition of the book appeared. The entire review is discussed in chapter 6, but it seems relevant here to focus on a few lines regarding the marketing of Hebrew books. After describing Hurwitz as someone he had met, the reviewer comments: "This man has traveled for the last ten years by way of our city and other cities in Germany and in other countries selling the fruit of his vision [his book]." If the review was written soon after 1797, this statement may refer to Hurwitz's wanderings well before the book was actually published. The reviewer then adds in a footnote:

> For this our heart grieves, over the abasement of the sages and writers of our faith because of our many sins. For the wealthy donors who also love and buy new books do not request them from booksellers but seek them directly from the authors themselves so that the latter go begging from door to door like peddlers bringing the first fruits of their thoughts to everyone's house. Subsequently, the number of book dealers has decreased and the dishonor of the Torah and learning has increased. It is also an embarrassment for us in the eyes of other nations who very much honor their own scholars.[45]

Was the reviewer referring disparagingly to Hurwitz himself, his intimate contacts with the rich and powerful, and the way he aggressively marketed his book, as substantiated by the letter to Fischer and the list of his agents?

Despite this reviewer's sniping comment, Hurwitz's amazing success was a source of enormous pride to him. Not only had he written a good book; he had also learned how to conquer the market! Thus he opened his second introduction to the newly revised version of his book, published in 1806–7 in Zolkiew by Abraham Judah Meir Hapfer, with a tone of enormous satisfaction:

> Blessed O Lord, God of Israel, from this world until the next, who has supported my soul to compose this composition and to publish it for the first time in Brünn in the state of Moravia in 1797. God allowed the beauty of the work to be seen by the eyes of all the dispersed of Judah, and it quickly spread throughout the entire world and was accepted with great honor in all places. Through God's desire, it appeared as far away as Mount Paran in Moslem lands, while its light [spread] on the wings of the earth and the islands of the sea so that it acquired fame in all countries. I then published two thousand copies, which circulated around the world. Besides Poland, Hungary, Germany, Holland, and England, it reached as far as France, Italy, the land of Uẓ, Damascus, and Aram, Algeria and Barberry, and Jerusalem . . .[46]

Hurwitz soon decided to republish the book in a greatly enlarged edition, since the first edition had already become scarce. But in the interim he learned the shocking news that the book had already been published by Joseph Rossmann in Brünn, one of his original publishers, without his permission and with the seven rabbinic approbations omitted. His response:

> You the truthful ones should judge these people and not show favor in judgment: the gentile from the city of Brünn with whom I originally published

> my *Sefer ha-Brit* in 1797 and his two Jewish advisors who worked with him in his craft. It is obvious that the gentile can do nothing and can certainly not publish a Hebrew book without the advice of these two Jews. Since he cannot read or understand the holy language and is ill equipped to decide on this matter, he naturally would ask them regarding [the propriety] of publishing any book. He would ask them their advice whether to publish this book or not, and their advice would be heeded. These three people thus sinned in publishing the first part of the book without my knowledge soon after I left to sell my book. I was impoverished, overcome with fear on my way, while they defied all the prohibitions of the seven supervisors of the community, the great rabbis who wrote and signed their names on *Sefer ha-Brit* in the city of Brünn ordering that the book not be published for a period of fifteen years from the time it was first published. . . . More than the damage they inflicted on me, they inflicted [it on] the entire Jewish community, since they stole the heart of every Jew in writing at the end: "This ends the words of the book." When in reality they had only published half the book, that is, the first part alone . . . but even in this part they made omissions in several places . . . and in one place I counted fifteen consecutive lines that they had left out.[47]

Thus Hurwitz expresses his bitterness regarding not only the theft but also the way the book was produced, as if the first part was complete in itself. He complains as well about the quality of the paper, the small print, the missing *haskamot*, and the dishonesty of claiming to produce a complete book when it is incomplete. He blames not only the publisher but also the Jewish proofreader, who should have refused to allow the "gentile" to publish and sell to Jews so damaged and illicit a book.

In the light of this bitter experience, Hurwitz pledges to produce a new edition of his book, this time with his name proudly displayed in both Hebrew and the vernacular. He warns future publishers and readers not to consider his book worthy unless it was complete, with no page omitted. He then issues a set of "covenants" outlining how his book should be published, formatted, printed, and read. The text is a precious document in the history of Hebrew printing and deserves to be cited in full.[48] It demonstrates the extent to which Hurwitz had not only acquired the business instinct to sell his book but also mastered the intricacies of publishing, proofreading, and printing.

Hurwitz's twelve covenants offer a remarkable glimpse into the workings of an early modern print shop and of the book industry of his day. They also reveal a deep insight into what the author expected of his readers—how he hoped they would read his book from cover to cover. In the end, Hurwitz was as demanding of the publisher, printer, proofreader, and reader as he was of himself.

Hurwitz begins this section by rescinding the previous ban on the printing of the book and announcing that from 1809 on, the book could be published by anyone without restriction. This permission is offered with the proviso that any new edition will be based on the new revised version. By this he apparently means the one recently published in Zolkiew in 1806–7. He then goes on to outline his twelve covenants. The first four are straightforward, as he insists that the print be attractive and readable, the font large or midsize but not too small; that the paper be of durable quality; and that the book be published in its entirety. He then makes the following request:

> The end of the first part [of the book] should be published on the same page with the beginning of the second part. The end of part one should conclude in the first column and the second part should begin on the second column of the same page so that they will be united and not separated. One part should not be separated from the other, not the first part from the second or the second from the first. There is no part that can stand on its own as in the case of the parts of other books. Each part of this book is not like the others, since I carefully preserved the order of God's redemption in it and it is connected from the beginning of the first part until the end of the second; every discourse is fastened to the preceding one and to the one next to it, and similarly each chapter. And thus all the words of the second part are dependent and rest on the words of the first part, and without part one, enlightened wise men will not comprehend anything. The first part is only an introduction to the second part, and it alone represents a gate to the house. A person who possesses only the first part is a gate keeper not a house owner. So in what way is it possible to divide this book? Therefore [the two parts] will be attached together on one page and the connection will be strong so that they will be stitched together as a permanent possession in order that the people will stand in a perfect covenant.[49]

To this unusual demand he adds that no index should be included in the book. He justifies this request on two grounds: First, because the different topics in the book are scattered throughout and not concentrated in any one place. The assumption that the first part deals exclusively with human wisdom while the second treats only the divine is misconstrued, he points out: "Words of Torah and fear of Heaven" can be found in the first part, while "human wisdom" is also in the second. His point is that the reader cannot read his book partially or haphazardly; it must be read from cover to cover without skipping anything. The subjects of the book are literally bound together as a unified whole that cannot be appreciated without this comprehensive examination of the entire work. While Hurwitz has a second reason for impeding selective reading of the book, he claims it is a secret and will not divulge it.

Hurwitz's peculiar requests seem to belie his claim at the beginning of this same introduction to *Sefer ha-Brit* "that all the wise men of Ashkenaz and the scholars of Berlin called this book an encyclopedia, that is, a book that gathers together all disciplines and all natural, mathematical, and divine wisdom and anything that comes to the mind of a human being that he wishes to know. . . . Almost everything is to be found in this book."[50] But an encyclopedia, at least in the modern sense, implies a casual, sporadic, or partial reading to quickly gain information on a specific topic isolated conveniently from all the rest. Clearly, this is not what Hurwitz meant by the term. His claim that one column cannot be separated from another, that both parts of his book constitute an uninterrupted whole, and that an index would defeat the purpose of continuous reading suggests that he did not mean to describe this modern form of encyclopedia but something entirely different. In contrast to what the "scholars of Berlin" might have conceived, in the tradition of Diderot and Voltaire and the encyclopedia project of the Enlightenment, Hurwitz had in mind a composition more reminiscent of the medieval or early modern encyclopedia, a kind of spiritual journey or ascent where the reader would be carried to a higher state of religious consciousness by the continuous reading that the composition required of him. This seems to be the meaning of his words "since I carefully preserved the order of God's redemption in it," implying that his alleged commentary on the prophetic/messianic tome of Ḥayyim Vital was itself messianic at its core.[51] By mysteriously concealing his second reason for justifying the continuous reading of his book, Hurwitz also betrays the esoteric nature of the composition. It should be perused neither casually nor sporadically but only with the aspiration to be illumined and transformed by it.[52]

Hurwitz's remaining six covenants deal with specifics of book production, known only by someone who had worked in or at least frequented a print shop. To begin with, he insists that the book be printed in quarto (that is, with four pages printed on each side of a full sheet, so eight pages per sheet; then folded twice to make four leaves or eight pages), though he is willing to consider octavo (eight pages per side, so sixteen in all, folded three times) as well. He will not accept the printing of abbreviations in his book under any circumstances, and he even presents his demand in the form of a religious injunction: "Any member of the Jewish community who . . . will remove any abbreviations which are already in this book and will sacrifice perfect 'burnt-offerings' by making all the words complete in the book of the covenant will certainly know that God will pay his reward and the work of his hands will be desired."[53]

He is equally insistent that the final letters of words not be dropped off at the end of a line: "One is forbidden in this book to draw a meaningless horizontal line [*kav tohu*] above the word indicating that the last letter is missing, as is the custom in other books." He singles out the "*Setzer* [compositor, typesetter] who attaches all the letters to the words so that all of them will be perfect. It is occasionally his custom to remove at will the last letter of the last word to make the work easier so as not to spoil the line but to align it a second time in an appropriate manner." But Hurwitz will have none of this practice, since it "destroys and soils the holy books, the divine words of eternal life, because in the heart of every Jew is the love of perfection." The *Setzer* should also make the ends of lines perfectly straight, since "it is often the custom of the typesetter who does as he pleases to fill out an empty space at the end of a line with an additional letter to straighten the line, filling the empty space with a letter that will begin the next line below it." Playing on the Hebrew word *davek* (meaning both to cling and to typeset), Hurwitz dramatically writes: "And you who cling to the Lord your God, I swear to you if your hearts are aroused to add even one abbreviation or to leave off the last letter of a word in this book and to put in its place a line above to indicate the missing letter or to fill the empty space at the end of a line with a letter that begins the next line, please be careful with such things in this book."[54]

Hurwitz next addresses the matter of pulling the lever to press the type on the paper, or *pressen zieher*. Having witnessed the process, Hurwitz laments the "pressers and pushers, most of whom are slothful and will not press with full human strength [so as to make a good, solid impression on the paper]. Subsequently the letters are not properly absorbed with ink on the book, especially if the paper is fit for writing called *Schreibpapier*." Thus the letters are not well recognizable because they are filled with bright white spots. The result is what Hurwitz calls *refa'im* (literally, ghosts), or weak impressions on the paper. Consequently, the printer should carefully supervise his pressmen, ordering them to pull hard with a strong hand so that the paper can fully absorb the inked letters.

Hurwitz concludes his twelve covenants with, first, a plea to proofreaders to concentrate when reading so as to avoid wandering thoughts and to refrain from speaking to others while performing their important work. He closes with a final admonition: "As long as I am living on this earth, no one should make a shortened version of this book or a section of it. He should never produce a single discourse or part of it to stand alone. One who shortens it will shorten his life, and one who divides it will divide his years [of life]. The most important point is that it should be proofread very well."[55]

It would appear that by the time Hurwitz wrote these lines for the new introduction to his expanded edition of *Sefer ha-Brit*, he had returned to Cracow, where he spent the remainder of his life. In the edition of the book published by Abraham Judah Meir Hapfer in Zolkiew in 1806–7, a notice appears at the end stating that the book can be acquired from the author, who currently resides in Cracow.[56] While Hurwitz also refers to being in Vilna and recalls again a visit to Buczacz, it is impossible to date these visits.[57] His gravestone records that he died in Cracow in 1821.[58]

There remains one more postscript to add to the account of Hurwitz's life and the life of his publications. In 1889, Joseph Fischer, a well-known printer of Hebrew books in Cracow, published a new edition of Emanuel Ḥai Riki's introduction to Lurianic kabbalah called *Mishnat Ḥakhamim* (The Teaching of the Sages), together with a commentary by Pinḥas Hurwitz called *Ta'am Ezo* (The Reason of His Counsel). That Hurwitz had always been interested in Lurianic kabbalah and especially its Italian commentators is evident from perusing the pages of *Sefer ha-Brit*.[59] Hurwitz had also mentioned in *Sefer ha-Brit* several other books he had written, including this particular composition.[60] Now, sixty-eight years after his death, he was honored with another publication produced in the same city where he had spent most of his life.[61]

Fischer announced at the opening of the book that he had bought this manuscript, together with another Hurwitz composition called *Mizvot Tovim* (Good Commandments), from a woman named Shifra Eichhorn. The next page contains three rabbinic *haskamot*, written by David Halberstamm, Solomon Halberstamm, and Akiva Karnitser. Karnitser appears to have been the moving force behind these approbations, along with another rabbi, Abraham Linzig, who prepared the manuscript for publication.[62]

The text then offers a short introduction penned by Hurwitz himself. Here he returns to the conditions he expects any publisher to follow when printing his book. He insists on large letters in Rashi script, recalling that one printer had rendered the text of *Sefer ha-Brit* in small letters, which strongly displeased him. To the reader of his second introduction to *Sefer ha-Brit,* his insistence that the book be published on good paper with no abbreviations is quite familiar. In addition, he requests that his commentary appear alongside the Riki text and that a list of errors be published at the end of the book. He concludes with his reasoning for writing a commentary on Riki's work in a statement that is reminiscent of his earlier justification for *Sefer ha-Brit* as a mere commentary to Ḥayyim Vital's *Sha'arei Kedushah.* Given the enormous interest in Luria in the author's day, and the importance of Riki's work as a summary of that kabbalist's approach, there is a need, Hurwitz argues, for a simple commentary to unpack that dense work and make it accessible to a larger audience. But unlike Hurwitz's more popular work, for which Vital's work was merely a pretext for exploring the natural world and more, this later work does not stray far from Riki's handbook. It is exclusively a kabbalistic commentary, no more.

In a manner that Hurwitz might have praised had he been alive, Fischer closes this introductory section by expressing satisfaction at having succeeded in publishing this useful work for beginners that the author had so wanted to see in print. He appeals to the potential readers of the book, especially in Cracow, where Hurwitz had lived, to purchase the book, which will allow the publisher to recover his initial investment and then to publish the other manuscript of Hurwitz's in his possession. Unfortunately, Fischer's hopes were not realized; this was the only time another book by Hurwitz would be published. Despite the continual editions and sales of *Sefer ha-Brit,* no other of his books would command a large reading audience. In the end, his fame rested solely and exclusively on his encyclopedia that he called "The Book of the Covenant."

CHAPTER 3

Why Should a Kabbalist Care about the Natural World?

THE MEANING OF SCIENTIFIC KNOWLEDGE FOR PINḤAS HURWITZ

ALTHOUGH Pinḥas Hurwitz had other intellectual interests, his primary passion was undoubtedly kabbalah. This is apparent, first and foremost, in his literary output, of which *Sefer ha-Brit* is by far his most important work. While the latter is encyclopedic in its scope, the author emphasizes that its primary objective was to serve as a commentary on Ḥayyim Vital's *Sha'arei Kedushah,* the popular mystical manual on imbibing the Holy Spirit. His only other published work, as we have seen, was *Ta'am Eẓo,* a commentary on Emanuel Ḥai Riki's introduction to Lurianic kabbalah, *Mishnat Ḥakhamim.* The other works he cites were apparently never printed but also dealt with kabbalah: a treatise on the mystical meaning of the commandments; another on the five secrets in the Book of Daniel; a commentary on *Sefer Yeẓirah;* and some notes on the prophetic kabbalist Abraham Abulafia. If one adds to this body of work Hurwitz's numerous citations of kabbalistic books in *Sefer ha-Brit,* especially those by the Italian interpreters of Isaac Luria's doctrines—Joseph Delmedigo, Joseph Ergas, Abraham Herrara, Sar Shalom Basilea, and Emanuel Ḥai Riki—his consistent fascination with this particular strain of mystical doctrine, namely that emanating from the Lurianic school, is apparent.[1]

Most of the first part of *Sefer ha-Brit* focuses on nonkabbalistic matters, primarily descriptions of the natural world, but the second part is of a different sort, meant to function as the culminating discourse of the book and consisting of an eloquent mixture of moral instruction, spiritual concentration, and mystical prophecy.[2] Although one might question the seemingly artificial connection between *Sha'arei Kedushah* and Hurwitz's book with respect to the first part of his composition, the inspiration and influence that Vital's work exerts on the second part is undeniable and profound.[3]

This intimate connection between the two works is especially reflected in a rabbinic quotation that both authors enthusiastically cite. Written in the name of the fourth-century rabbi Pinḥas ben Yair, it speaks of his belief in human perfectibility and the steps required to receive the Holy Spirit. The Law, he claims, "leads to carefulness; carefulness, to diligence; diligence, to cleanliness; cleanliness, to isolation; isolation, to purity; purity, to piety; piety, to humility; humility, to fear of sin; fear of sin, to holiness; holiness, to the reception of the Holy Spirit; and the Holy Spirit, to resurrection."[4] Vital surely saw this text as supporting the major claim of his book, which was that by following a step-by-step spiritual journey one could ultimately connect with the divine.[5] Hurwitz, meanwhile, not only underscored the importance of this text in and of itself, but he also transparently personalized his connection with the name of its author. After citing Vital several times, he mentions in the same breath the love of humanity displayed by his ancestor Pinḥas ben Yair and his own desire [i.e., that of Pinḥas ben Eliyahu] to make peace in the world.[6]

Hurwitz emphasized on several occasions his identification with Vital's position that while prophecy was geographically and temporally restricted, imbibing the Holy Spirit "is not restricted to a specific place or time and is accessible both in the land of Israel and outside of it in every generation for eternity."[7] In fact the opportunity for spiritual fulfillment was greatly enhanced in his own generation by the availability of the printed books of Lurianic kabbalah, which had been unavailable previously. Indeed, in this age, Hurwitz wrote, "the gates of light and mercy were opened, for it was closed to the time of the end of days, the happiness of the commandment, and the great present joy of God to publicize the honor of the name of his kingdom forever and ever. In particular, at this time all the holy writings of Isaac Luria were published who opened up for us the gates of Torah, which were closed and sealed with a thousand clasps from ancient days, and all his words are the words of the living God as related by Elijah the prophet."[8]

Armed with this knowledge of Lurianic kabbalah and its commentators, Hurwitz proceeded to elaborate and expand upon the specific instructions Vital had offered, especially in the third book of *Sha'arei Kedushah*, in preparing for the spiritual journey uniting the soul of the believer with the divine. Here, for example, are his directions for creating an atmosphere of spiritual solitude and purification, which would allow the ultimate unification to take place:

> One should confess, and then immerse [in a *mikveh,* or ritual bath]. After this, he should seclude himself in a room where he will not hear even the sound of birds chirping, and all the more so human voices, so that he will not be distracted. If it is possible to do so after midnight, this is even better. He should light many candles; and if [he wishes to meditate] by day, the best time is before noon, while garbed in a *tallit* [prayer shawl] and *tefillin* [phylacteries]. He should close his eyes and remove his thoughts from all mundane matters, as if he no longer existed in this world. Afterward, he should begin to sing praises to God from the praises of David [i.e., Book of Psalms] with great fervor. Then he should contemplate the supernal worlds, the hidden lofty levels, from below to above, and picture the upper worlds in his imagination; he should begin to imagine that his higher soul is ascending higher and higher according to its soul roots [*shorshei neshamot*] in the "man of souls" [*adam ha-neshamot*] of the heavens of the World [or level of reality] known as "action" [*asiyah*].[9]

The rest of the passage is filled with the technical vocabulary of Lurianic kabbalah and is of interest only to the initiated. But this citation is sufficient to indicate Hurwitz's seriousness in offering a detailed and practical guide to moral perfection, spiritual elevation, and the ultimate serenity attainable in connecting to the divine. At the very end of his volume, Hurwitz again underscores the fact that *Sefer ha-Brit* is ultimately a practical guide to spiritual illumination:

> Examine, my brother, the main point, which is not to dwell on this chapter and discourse simply in order to understand or teach it, but please read it in order to instruct yourself and to accustom your soul to do everything that is written here. Make it prominent so as to practice it in order to know how to succeed in the great matters we have mentioned. Place it always on your forehead in all its manifestations and regarding all that has been said about it in your dwelling, in your home, and in your going by the way, in your slumber and in your rising up. Thus you will see if the chimneys of the heavens will open to you as well as the gates of light and awe, of holiness and love, and of happiness. Then you will be transformed into another per-

> son and you will suddenly acquire the Holy Spirit from the Holy God. You will then proclaim that the thing I heard was true but I did not believe it until I was tested. And behold I was not even told half of all this wonderful goodness. It is simply joyous worship, which is unlike anything else. God will privilege us to be those who cling to the divine so as to experience the joys, happiness, and great love all the days of our lives, as the biblical verse states: "And you who cling to the Lord you God all of you are alive today" (Deuteronomy 4:4).[10]

At the very least, these small selections gleaned from numerous pages of moral suasion and spiritual guidance offer some indication of the culmination of a journey that began with the natural world but eventually left it behind in pursuit of supernal mysteries and the ultimate coadunation with the divine realm. While these spiritual instructions are not the primary focus of my own inquiries into Hurwitz's text, the results of which follow in this and the next chapters, neither should they be ignored. On the contrary, they stand behind every aspect of *Sefer ha-Brit*. In the end, as Hurwitz aptly puts it, you must read his book not merely to understand and teach it, but to let it transform you into another person!

HURWITZ AS SCIENTIFIC WRITER

If the mystical trajectory of the second part of *Sefer ha-Brit* and its intimate connection with *Sha'arei Kedushah* are apparent, what about the first part, that dealing primarily with the natural sciences? Was Hurwitz actually serious when he declared in the introduction that in order to properly interpret the kabbalistic ethical treatise of Ḥayyim Vital, particularly its third section, he needed to present a full accounting of the natural world?[11] How credible was his claim that all he was doing in hundreds of pages of scientific explication was elucidating Vital's actual meaning? Vital had little to say about the natural world in his short book, and nothing in Hurwitz's previous background intimated the following statement of his:

> I therefore arose to open the gates of holiness to my beloved with divine support as an entrance to the hall which had initially been only a small crack, as the author [Vital] himself testified. I said to myself, if God wants me to succeed, I will compose a book that will be a kind of introduction whereby the doorways will be opened, and a righteous nation will enter

the gates of holiness. I will open to you its floodgates of heaven so that its doors will no longer be shut. Thus I will bring all the necessary introductions to understand his words, and I will explain every place that requires an explanation, and where he made his words concise, I will expand them and complete them so that every person will understand and penetrate his words.[12]

Thus apparently convinced that he was merely "filling in the blanks," so to speak, of Vital's modest composition, Hurwitz proceeded, in the largest part of his big book, to offer a systematic exposition of the following topics: astronomy and cosmology, the elements, geography, meteorology, mineralogy, botany, zoology, embryology, anatomy, and psychology. For Hurwitz, there was no doubt that knowledge of the natural world was propaedeutic to the human ascent to imbibe the Holy Spirit and pass through "the holy gates." But why, one might ask, was the one contingent on the other? Why should a kabbalist bother to master natural philosophy when his ultimate goal was to transcend his earthly existence and enter a spiritual place beyond nature itself? To the reader of *Sefer ha-Brit,* this is not a casual question; indeed, it offers the key in understanding the ultimate purpose of the book and the author's primary message to his Jewish readers.

Those few scholars who have investigated the scientific sections of Hurwitz's book have focused primarily on his sources, both Hebrew ones and those written in other European languages, primarily German, the only other language Hurwitz presumably knew to some extent besides Yiddish. Such studies, which are only partial, have yielded neither dramatic nor unexpected results. Hurwitz was acquainted with and cited much of the Hebrew library of scientific works composed by the end of the eighteenth century. These included the writings of Barukh Lindau, Tobias Cohen, Israel Zamosh, David Gans, Mordechai Shnaber Levison, Moses Ḥefeẓ, Joseph Delmedigo, Abraham Herrera, Barukh Schick, and more.[13] As mentioned in the previous chapter, he was especially reliant on Lindau's *Reshit Limmudim,* with which he sometimes disagreed but nevertheless readily copied, often without indicating his source.

The only scholar who has seriously looked beyond his Hebrew sources is Resianne Fontaine. In an initial probe of his sources on zoology and meteorology, she offers some tentative suggestions regarding other scientific authors Hurwitz might have known. They include George Christian Raff; George-Louis Leclerc, comte de Buffon; Johann Christian Polykarp

Erxleben; Petrus van Musschenbroeck; and others.[14] Undoubtedly, similar close comparisons of Hurwitz's text with other contemporary sources, both those he acknowledged and those he did not, will allow us to understand precisely how he composed his compendium of scientific knowledge. While such studies are surely necessary to fully understand Hurwitz's originality as a disseminator of scientific knowledge, I doubt whether they will significantly alter our present assessment of him as a student of the sciences. It is already clear that, despite his prodigious efforts, his treatment of the sciences was mostly unoriginal and conventional. As he himself made clear, he had neither formal university education nor medical training, nor was he intensively exposed to systematic instruction or vast library resources. He was an autodidact whose knowledge rarely transcended the limitations of his traditional education.

Yet despite these constraints, his accomplishment as a scientific educator was enormous. His book surpassed in popularity all the other books on which he relied. He had obviously found a formula for teaching the sciences to traditional Jewish readers that no other author had mastered as well as him. The questions the rest of this chapter poses, then, are, to my mind, the most significant one might ask, to wit: Why was science so critically important to Hurwitz? Why did he believe it was connected directly to the process of mystical ascent and prophecy? And why did his understanding resonate so thoroughly for his numerous readers who found his depiction of the sciences the most meaningful and appealing of all?

THE NATURE OF SCIENTIFIC KNOWLEDGE

To attempt to answer these questions, we might begin by focusing on a theme raised several times in Hurwitz's large book: the ability of human beings to know and understand the world. In a discussion on whether or not there are four elements, Hurwitz acknowledged

> the great confusion among modern scholars regarding a matter well known, received, and accepted from earliest times until now by all ancient scholars, both Jewish and non-Jewish, that there are four elements and their ordered relation to each other. Now there are scholars who claim there are only three elements, those who declare there are two, and still others who assume there is only one, and each [hypothesis] is of equal value. . . . But don't say that the ancients were not as smart as the moderns, because

> the mind of the ancients is like the entrance to the [heavenly] palace . . . and their opinion is wider than ours. It is only that experience has come to deny their words and falsify their opinions. And what has happened to the ancients will also happen to the scholars of our generations, so that the moderns will ultimately not be happy. Then in the end of days, discoveries that are still hidden from us will be revealed through the invention of new instruments, and these will contradict the assumptions of the scholars of our own generation. Those future scholars will no longer remember favorably the views of our contemporary ones, for in another generation all they accepted will be erased. Their sons will rise after them to discover even better instruments and clearer experiments to deny their former conclusions. Everything they had accepted will be destroyed by propositions based on philosophical logic resting on the wings of rational human assumptions. . . . Thus all external sciences are far from the truth, with the exception of mathematics and geometry. Everything else will be swept away in a wind of vanity, since it has no force to establish the truth.[15]

At this point, Hurwitz turned to refute his contemporary Solomon Maimon, who had claimed, in the opening to his commentary on Maimonides's *Guide for the Perplexed*, that philosophy in his day had reached a state of perfection. The absurdity of this position, Hurwitz said, was made clear by Kant's frontal attack on philosophy, which confirmed dramatically for Hurwitz the accuracy of his position.[16] Leaving aside for the time being Hurwitz's attack on Maimon and his enlistment of Kant to bolster his epistemological position, let us define more precisely Hurwitz's understanding of science and knowledge. For him, all human knowledge was finite, tentative, and time-bound. What one generation knows, the next will overturn: "All the scholars of this last generation arose and contradicted the opinion of recent scholars, since it is their foolish way that one builds while the other demolishes; one dreams while the other solves."[17]

In sum, Hurwitz understood precisely the nature of scientific knowledge and the methodology of the scientist. These rested on problem-solving, the accumulation of new data and new observations, and the constant reformulation and readjustment of scientific understanding to accommodate the unlimited flow of new and contradictory information. This strong empiricism came even to challenge the authority of Judaism's greatest philosopher, the vaunted Moses Maimonides. In a later discussion of the nature of prophecy, a subject critical to the project of interpreting Ḥayyim Vital's small book, Hurwitz strongly objected to Maimonides's definition of the prophet as a person possessing perfect knowledge:

> He [Maimonides] claimed that prophecy only rests on a courageous and rich wise man, one who possesses all rational faculties. He meant by this that a wise person is one who knows all rational things, all science, all of Torah and every discipline. But such a thing is impossible for human nature, since there is no end to all knowledge and wisdom, especially in the case of Torah learning, whose length is unfathomable. Thus when the biblical verse speaks of "the person who is smarter than any other person" [cf. I Kings 4:31], it also proclaims "I said I will get wise but it is distant" [Ecclesiastes 7:23].[18]

If complete knowledge were a condition for prophecy, Hurwitz added, prophecy would be impossible to attain, since each generation discovers its own knowledge unavailable to the previous one. Human knowledge is always relative, such that a person who is intelligent is defined in reference to other scholars of his generation, never in absolute terms.

In contrast to the medieval philosopher lionized by Maimonides who knew the world through his rational deductions, the true scholar, for Hurwitz, learns inductively from his experiences, for "there is no wise man like the empiricist." And given the ability of human beings to know but partially and sporadically, only a knowledge based on faith offers them the certitude and security of full comprehension of the world and its divine plan. In the words of Habakkuk 2:4, "The righteous lives by his faith," to which Hurwitz adds: "by his faith and not by his human investigation."[19]

THE ALLURE OF SCIENTIFIC DISCOVERIES

In the light of Hurwitz's unwavering fideism and confidence in the tradition of his forefathers over any human constructions of knowledge, let us now return to the question we raised earlier: Why was Hurwitz attracted to science in the first place—to describing the natural world to his readers, devoting hundreds of pages to descriptions of animals, mineral, plants, and meteors—if he had so little trust in the human capacity to understand the world and recognized its utter fallibility in the absence of faith based on divine revelation? Why was nature study, as opposed to philosophy, a prerequisite for reaching the level of prophecy? The answer to these questions might now be clearer. Hurwitz was attracted to science primarily because it documented the ongoing process of discovery. It appealed to him neither for its rationality nor for its systemic order; on the contrary, it was its confusion, its inconsistencies, its haphazard nature, and its constant reversals that excited Hurwitz as well as his readers. Expressed differently, the history of scientific discoveries, in overturning conventional and long-held assumptions about the world, served to reinforce the miraculous, the supernatural, and the divine hand in creation.

Perusing Hurwitz's handbook of scientific information, the casual reader will easily notice what delighted Hurwitz the most and led to his most elegant and eloquent descriptions: namely, the new scientific discoveries of his day. They range from the barometer and air pump to the helium balloon, the diver's bell, the lightning rod, and remarkable sound machines. Some of these items, such as the barometer and air pump, were fully conventional by Hurwitz's time, and his descriptions were lifted from other Hebrew works.[20] But some of his descriptions appear more original and seem to be based on information gleaned from either written or oral sources known to him alone. Take, for example, his description of Paul-Pierre Blanchard's flight in a helium balloon:

> The scholars recently invented an instrument called a luft balloon whereby human beings ascend and descend into the atmosphere to roam . . . and with it they fly to the heavens, and accordingly, a human being can fly to the top of the atmosphere to the face of the firmament playing like a horse and rider; and what can a person do after flying in heaven like an eagle? And out of my great love for you, dear reader, I declared: "Can I hide such a wonderful thing from my beloved? It is certainly not right to do so, and thus I will not restrain my pen from writing every line regarding this instrument and its construction and the reason for everything in it until one understands and fathoms so that he can do everything to build such an invention."[21]

Hurwitz then offered detailed instructions on the construction of a hot air balloon, specifying the use of a large bag called an envelope that contained heated air, with a gondola or basket suspended beneath it. At the end of his long description he mentioned Blanchard directly: "For who ascended to the heavens and descended such as the Frenchman named Blanchard, who rose and came down many times while many people watched." He related how Blanchard once landed in a field in Holland and the farmers beat him, thinking he was a heavenly creature; and how he flew thirty-eight times, including a historical flight across the English Channel in four or five hours in 1783 or 1784, traveling from Dover in England to Calles (Calais) in France. He concluded: "How great are the acts of God, the fashioner of man and the creator of wind. You have made him master over your handiwork, laying the world at his feet through your acts [Psalm 8:7]. For God bestows his spirit into man's hands to be a traveler over seas and lands. How magnificent is your name in all the earth!"[22]

Hurwitz's emotions were similarly stirred by the recently invented diving bell, first conceived by Edmond Halley in 1691 but actually designed and implemented by the English engineer John Smeaton in 1789:

> Know, my brothers, how great is the joy of European scholars of the last generation who discovered with their vast knowledge this aforementioned invention by which a man descends into the heart of the sea and the depths of rivers. The source of this device and its construction is based on their deep investigations of natural philosophy. They conceived that water is unable to enter a closed air space. . . . They were overtaken with joy and happiness by this discovery and thus declared: Who is like us who teaches wisdom and the wonders of the secrets of nature as we do today. There is no one like us among all the nations around us, including the pathetic Jews who dwell in our midst. They are a nation lacking intelligence except in the Talmud in their possession, and the Talmudists themselves have not mastered any discipline.[23]

Hurwitz's exhilaration over the new invention was tempered by a feeling of inadequacy and cultural inferiority, in that Christians had mastered the secrets of nature while the impoverished Jews stood on the sidelines of contemporary culture, incapable of any such production. Sensing the need to salvage the Jewish reputation for inventiveness, he cited a rabbinic text which suggested that the rabbis had long known how to invent such a machine.[24] However far-fetched this notion was, it allowed him to blatantly declare:

> How great is the wisdom of the rabbis in all scientific disciplines, and how deeply they penetrated the science of nature. . . . Thus the recent scholars should not rejoice or claim that they were wholly original, for this was a small matter for the rabbis of the Mishnah and the Talmud given the expansiveness of their minds in wisdom and intelligence, knowing every

small and great thing. And thus, my brothers and people, may your hearts live forever and may you be joyous in all that your hands touch for the sake of the Talmud, received and accepted by a people chosen for its wisdom among all the nations![25]

These two examples of Hurwitz's descriptions of dramatic discoveries suggest his great enthusiasm for the achievements of inventors and discoverers of his own generation and the special prominence he afforded them in his summary of the sciences.[26] In the case of both the hot air balloon and the diving bell, he felt moved to express deeply religious feelings; in his view, these inventions offered testimony to God's beneficence on earth and the miraculous glory of his creation. In the case of the diving bell, his reportage also afforded an outlet to defend the honor of the rabbis, who, in his estimation, had been slighted by the special attention given to Christian inventors. That he chose to see the discovery along ethnic-cultural lines of superiority and inferiority is surely a reflection of his sensitivity to the low cultural profile of the Jewish community in his era and might be linked to some of the criticisms he voiced later in the book about the social conditions of contemporary Jewish life.[27] Interestingly, however, his effort to elevate the learning of the rabbis stands in contrast to his stronger argument (to be discussed in the next chapter) that the origin of philosophy and, by implication, science came not from Jerusalem but from Athens and Rome. By feeling compelled to show that the rabbis were actually precursors of the latest trends in science, he appeared to contradict himself.

The primary function of these elaborate narratives of scientific discovery was not to preserve the honor of the rabbis, however, or even to convey a sense of wonderment and acknowledge the majesty of God and his creation for Hurwitz and his readers. They did something else even more critical for the author: they upset and confused the supposedly ordered, stable human systems of knowledge based on long-standing rational assumptions and intense scrutiny of nature by philosophers and scientists. In their chance occurrences and in their indifference to the rules and norms by which nature was expected to function, the new discoveries appeared to undermine the possibility that human beings can fully understand anything. In short, the very ability of human intelligence to grasp the complexity of God's handiwork was called into question.

For the author of *Sefer ha-Brit*, God was wondrous, and the new discoveries that seemed to occur daily reconfirmed the miraculous and supernatural dimension of the material world. The empirical study of natural processes was intellectually compelling, for it underscored the limits of human knowledge and experience to appreciate fully the multifariousness of creation. It was also religiously compelling in bolstering the sense of awe necessary for an unquestioning faith in God. Nature study was for Hurwitz the ultimate resource in imbibing the Divine Spirit, since it revealed constantly the majesty of the creator and the finitude of his created beings. This insight ultimately led to the peace of mind and security of the true believer. And when the investigation of nature was divorced from the injurious pursuit of philosophical metaphysics, as Kant had recently pronounced, the road was paved for a pure belief untainted by doubt and heresy, enhanced by kabbalistic sapience, and worthy of prophetic inspiration. Armed with this understanding of nature and its relation to faith, Hurwitz chose to conjoin the study of the natural world with that of the kabbalah, and to link the inspiring stories of novel inventions and discoveries to the moral and spiritual enhancement of his imagined readers.

EDWARD JENNER AND THE VALUE OF SMALLPOX VACCINATIONS

The most impressive and original description of a recent scientific discovery in *Sefer ha-Brit* was surely that of Edward Jenner's smallpox vaccine. The history of smallpox inoculation and cowpox vaccination has long been considered by historians of medicine and public health as a significant development in the emergence of modern medicine and in the authority it was gaining throughout the Western world. The modest beginnings of inoculation (with live smallpox virus) at the beginning of the eighteenth century and Jenner's dramatic discovery of the much safer procedure of vaccination (with the related cowpox virus) by its end offer remarkable testimony to the impact of scientific discoveries on state policy and the shaping of a new public consciousness regarding health care. In a more general sense, the story of the fight against smallpox is also related to new attitudes toward religion and science, the changing role of the medical professional in relation to sacerdotal and political authority, and the newly sensed power of human beings to resist and even overcome any fatal condition. The new procedure was revolutionary even from the perspective of the history of medicine, since inoculation, and to a lesser extent vaccination, meant giving a disease to someone who was healthy, the very inverse of the traditional role of the doctor.[28]

I have elsewhere described several contemporaneous Jewish responses to the discovery of smallpox inoculation, including that of Hurwitz, in the larger context of the tensions between medical and rabbinic authority in the modern era.[29] In this chapter, I would like to consider again Hurwitz's depiction of Jenner's discovery and its impact on the Jewish community from the particular vantage point of the themes of this chapter. As we shall see, Hurwitz's fascination with Jenner's discovery was fully consistent with his approach to the other discoveries already described.

Hurwitz was not the first Jewish author to describe the smallpox vaccine, but his discussion was one of the most extensive. Whether or not he actually read Jenner's treatise in one of the many languages into which it was soon translated, he certainly displayed a detailed knowledge of Jenner's discovery, the procedures he followed, and the subsequent history of vaccination among both Jews and non-Jews. Hurwitz opened with a description of smallpox, its various names, its horrendous effects, particularly the scarring and pock marks it left on its victims. He mentioned in passing its possible origin in the impurity of menstrual blood. He reviewed carefully the process of inoculation, which, he noted, had been practiced for almost a hundred years. He emphasized that although inoculation had considerable success, it did not always produce the expected results, and some patients still died. That is why "the sages of Israel abstained from this procedure, not wanting to permit it for members of our community so as not to involve our children in a doubtful case of saving life from the outset." But this new discovery was of an entirely different sort for Hurwitz, and thus "it is a commandment to publicize this examined cure which God bestowed on us in this generation, that which previous generations were not privileged to enjoy."[30]

Hurwitz's primary motivation in spreading the good news was to assure his readers of the remarkable safety record the new vaccine had achieved: "Already thousands and ten thousands of people have tried it, and all of them came out of it safely, not even one dying." When the rabbis saw these results, they immediately offered their support: "When the sages of Israel saw that the procedure was examined and carefully tried and that this practice spreads day by day and is accepted in every state to the ends of the earth and to far-off islands, and no one has heard or seen any impediment or flaw affecting any person great or small, since it is a procedure which never harms anyone, they all arose, the elders of Israel and the supporters of Torah in this generation, and permitted it to all of Israel according to the Torah, commanding the doctors to do this to their children and grandchildren."[31]

Hurwitz then offered his readers a detailed description of Jenner's empirical studies of the milkers who, coming in contact with infected cowpox, immunized themselves from the more dangerous smallpox. He described Jenner's method of injecting the cowpox beneath the skin, the waiting period as the sore appears, opens, scabs, and falls off. Hurwitz noted that the scab should never be picked off and that one's hands should be kept clean. The immunization usually worked the first time, he observed, but occasionally needed to be repeated.[32]

Hurwitz listed four advantages of the cowpox vaccine versus smallpox inoculation: It was more accessible, more reliable, and less contagious; it could be focused on one area of the body alone; it did not spread easily to other parts of the body; and it was not dangerous in affecting children at the vulnerable times when their teeth were emerging or they were susceptible to other diseases. He cautioned that proper medical supervision was necessary to ensure good results and that the vaccine should be administered early in order to prevent the spread of smallpox from the outset.[33]

He again emphasized the reliability of the cure, stating that it had been universally accepted already for some thirty years among all peoples of the earth, including those on the American continent.[34] He reiterated that the practice of vaccination had begun to spread among Jews. This time he singled out a particular physician: "And the *halakha* is according to Rabbi Simon, doctor in the city of my residence, Cracow . . . in his pamphlet *Terufah Ḥadasha* [A New Cure], published in 1803–4, in which he urged the members of our community to give this tested medicine to all their seed, whether male or female, prior to hearing of or viewing the presence of this very dangerous natural smallpox. And thus the holy community of Cracow and others act accordingly. Simon the righteous actually showed me a list in his ledger of hundreds of children vaccinated by him, and all of them came out of it safely, perfect and unscathed on their bodies and all of their organs strong and firm."[35]

I have been unable to locate the booklet by Rabbi Simon, but the author can be easily identified. He is Dr. Szymon Samuelsohn (or Samuelson), born in Frankfurt an der Oder, where he also studied medicine between 1773 and 1779. On the recommendation of Count Sułkowski, he was invited to serve as court doctor in Cracow, where he was allowed to live in the Christian quarter of Kazimierz despite his Jewish ancestry. In 1781 he was appointed the physician of the Jewish quarter. He later settled in Warsaw and Lublin, where he served as court doctor of the king. Hurwitz obviously enjoyed a personal relationship with this respected doctor, who no doubt used his political status with Christian patrons and his official position in the Jewish community to encourage his co-religionists to be vaccinated.[36]

Hurwitz concluded his observations by stressing the novelty of the disease. He repeated that there is no mention of smallpox in all of Jewish literature, including Maimonides, nor in classic medical literature such as Hippocrates, nor even in the popular contemporary medical textbook of Tissot, which he had consulted.[37] He closed with a plea to Jewish parents to consult doctors and to prevent the danger of smallpox by getting their children vaccinated.[38] Clearly, Hurwitz considered preserving one's health by heeding the most up-to-date medical opinion a high religious priority.

MARCUS HERZ AND PINḤAS HURWITZ ON INOCULATION

Hurwitz's enthusiastic endorsement of smallpox vaccination is best appreciated by comparing it to the negative reaction of the German Jew Marcus Herz, distinguished physician, philosopher, and close associate of Moses Mendelssohn. Despite his deep commitment to enlightenment values and his own secularity, Herz strongly opposed the use of the Jenner vaccine for treating smallpox. Nearing the end of his life in 1801 and still fully committed to his understanding of "philosophical medicine," Herz objected to the use of "brutish" material to cure human illness, concluding that, philosophically and morally, the entire procedure was categorically improper. Even though this stance was condemned by his medical colleagues, significantly tarnishing his professional reputation, he refused to admit the simple fact that the vaccine worked.[39]

Soon after Herz had articulated his position on vaccination, one of his chief critics, Jason Ezekiel Aronsson, reminded him of his earlier and highly visible critique of the rabbinic establishment over its unyielding stand on early burial.[40] In 1787, Herz had pitted his rational moral commitment to relieving individual human suffering against the seemingly uncaring orthodox rabbinate, which was willing to ignore several incidents of live internment so as not to alter tradition and custom. How was it possible, Aronsson mockingly asked, given Herz's earlier high moral stance, that he should now display what amounted to a similar indifference to human suffering in the case of vaccination?[41]

Thanks to an insightful portrait of Herz by Martin Davies, it is possible to understand these inner contradictions. In the case of smallpox vaccination, Herz's conception of science was informed by his metaphysical interests, which would not allow him the openness to accept empirical insights derived from purely pragmatic procedures. When he published his polemic against Jenner in 1801 at the very end of his career, the philosophical premises informing his medical practice seemed to have been thrown into serious jeopardy. Inserting cowpox into human beings subverted the notion that man was the noblest creature on earth, in possession of a soul that empowered him with unique moral, intellectual, and physical attributes. Vaccination, rather, suggested a physiological affinity between human beings and animals, an idea that was an anathema to him and might, he feared, lead to the brutalization of humanity itself.

Not only that, but vaccination thoroughly upset Herz's rational construction of how illness was caused and how it should be treated. In his understanding, the logical character of illness could be discerned through a system of analogies, allowing an illness to be matched with a cure with seemingly reliable results. All of this was threatened by the sheerly accidental nature of the Gloucestershire milkers' discovery, which seemed to obviate the rational construction of science through logical analogies. Experimentation could not be limited by a closed system of prior assumptions; instead it had to be free to proceed from observation to observation if it was to have any practical value. For Herz, among other results, the vaccination controversy brought to the surface the great divide between idealistic philosophical medicine and scientific progress. In the words of Davies, Herz's final plea for his position, against all odds, was "one of the grand—even if paradoxical—gestures of the Enlightenment."[42]

In 1787, in the midst of the early burial controversy, Herz had challenged a rabbinic orthodoxy on the basis of a scientifically vindicated ethic and a commitment to medical progress. He had also attempted to redefine the social jurisdiction of the doctor in relation to the rabbi. Given the specialized knowledge of the doctor over the clergy, should not the latter defer to the former? The physician, armed with the cumulative knowledge of medical practitioners, was in a far better position to provide advice for the maintenance of human life. But seventeen years later, Herz was incapable of recognizing the emergence of a new approach to combating disease and the fact that medical progress was making his version of scientific theory obsolete. If science was to be true to itself, it had to rid itself of all orthodoxies, including its own. The doctor's authority, unlike that of the theologian or religious sage, no longer rested on a fixed system of prior assumptions but on empirically driven, open-ended reassessment of the physical world.

Hurwitz, however, with little investment in the rational underpinnings of science, was able to take the chance discovery of the milkers in stride. Having no stake in the philosophical physician's premise that human beings were distinct from animals and that the treatment of disease could be rationally constructed, Hurwitz had no qualms about overturning existing philosophical and scientific norms. The only thing that mattered was that vaccination had proven effective and lives were spared. Hurwitz's interest lay in questioning the reliability of human intelligence to make sense of the order of the universe and creation—a far cry from Herz's struggle to reconcile reason and empiricism. Medical and scientific breakthroughs that overturned rational human schemes were, for Hurwitz, the very essence of his commitment to studying science and informing his readers about its dramatic findings. It is thus no small irony that the kabbalist, rather than the philosopher, would not only endorse the controversial vaccination of Edward Jenner but also promote its immediate and regularized adoption within the Jewish communities of eastern Europe.

CHAPTER 4

Judaism and Metaphysics

HURWITZ'S EPISTEMOLOGICAL AND HISTORICAL CRITIQUE OF PHILOSOPHY

In contrast to his enthusiastic endorsement of the study of the natural sciences among Jews, Pinḥas Hurwitz strongly objected to the study of philosophy, specifically metaphysics. In his recent book *The Origins of Jewish Secularization in Eighteenth-Century Europe,* Shmuel Feiner singles out this aspect of Hurwitz's thinking, placing him squarely in the camp of orthodox Jews, although initially he tempers this assessment: "Because of the zeal with which Hurwitz defended the faith against the attackers," Feiner writes, "he ought to have belonged to that group of God-fearing orthodox Jews. But his broad horizons and his up-to-date knowledge in science and philosophy actually made him a suitable candidate for the camp of the *Maskilim.*" Yet despite this acknowledgment of Hurwitz's complex ambiguity, Feiner prefers in the end to brand the author of *Sefer ha-Brit* a member of "the congregation of believers"; considers the goal of his "orthodox" project as one of proving the philosophers wrong; and says of this "orthodox author" "that in the contest between science and faith, he unhesitantly chose faith."[1]

My goal in this chapter is to challenge and problematize this characterization. In the above passage, Feiner conflates science and philosophy, whereas for Hurwitz the study of physics and metaphysics were hardly identical, as we have seen. But besides this confusion, there is also the problem of equating an antiphilosophical position with orthodoxy. Because Feiner does not offer a precise definition of the term (though he uses it repeatedly), one might assume that by "orthodoxy" Feiner simply means a fundamentalist belief in God, his revelation, and the observance of the divine commandments. But is this what "orthodoxy" meant in its early-nineteenth-century context, and is it really incompatible with philosophy or rationality in general? Can a commitment to disseminating scientific knowledge and universalistic morality among Jews with a simultaneous distrust for metaphysics be reduced merely to an "orthodox" project and thus located within the camp of the "the congregation of believers"? Is such a position of "faith" inconsistent with rationality and science?[2] Hurwitz's highly nuanced position regarding the place of philosophy in Judaism, as well as the kinds of philosophical and historical arguments he employs to argue this position, requires more precise elucidation.

THE ALLEGED DECEPTIONS OF MOSES

Feiner is correct in acknowledging that Hurwitz knew a great deal about philosophical, scientific, and historical developments in his day. Although he was hardly a philosopher by training—he did not display a profound understanding of philosophical discourse, nor does it appear that he read deeply in the philosophical literature he cited—he was nonetheless well aware of general intellectual trends both within and beyond the Jewish community, and he used this knowledge skillfully to elaborate a consistent, well thought out position regarding the place of philosophy in Judaism, the relation between physics and metaphysics, the historical relationship between the Greek philosophical tradition and Judaism, and a definition of faith for Jews.

We begin by examining the key sections of Hurwitz's long chapter entitled "Derekh Emunah" (The Way of Faith), presenting them sequentially, the better to see how Hurwitz developed his multipronged argument regarding the relationship between philosophy and Judaism.[3] Early in the chapter, Hurwitz seeks to make clear to his reader that he is fully aware and up-to-date regarding the contemporary challenges posed by philosophers in his own era:

> You might think, dear reader, that the author was not really informed about the philosophical ideas of our time and the heretics of our generation, and that all that he wrote until this point is not sufficient to answer them, since they deny everything, claiming that the world is eternal and operates according to an unchanging nature and that the Infinite is too lofty to supervise this world and to know everything that happens under the sun. But I shall respond by declaring that I am more familiar with their words than you are. I also heard their opinions as you heard.[4]

To this deist view of God, he adds the view of the atomists and followers of Spinoza regarding the chance occurrence of creation and the eternality of the world. He then pauses to discuss a belief that all of these philosophers shared, namely, "that Moses never ascended on high and that God never descended on Mount Sinai."[5] He continues:

> They claim that Moses was a great scholar in natural philosophy and a skilled politician in ruling the state. His knowledge was greater than that of the Egyptians, and he accomplished everything based on his own knowledge, understanding, and great intelligence. He took a blind people out from Egypt not with a strong hand or with a sword and spear but with wisdom, since wisdom is better than instruments of war. And behold, a people exited Egypt with him who were accustomed to mortar and bricks, learned [only] in the work of the field upon whom the light of intelligence had not shined. They were a poor and impoverished nation who had acquired no sciences. Thus he accomplished on Sinai everything through machinations based on natural secrets that had not been known by anyone before him and through wonderful instruments that had not existed in former times. The people actually heard the sounds and saw the flames, although he had prepared them on the mountain by natural means and with instruments made up of gunpowder, now called *pulver*, and a barrel of fire called a *kanon*. People heard his words in the midst of the fire, cloud, and fog around him created by the *pulver*. From there he spoke to them from a distance by means of a sound device called a *tubus lactarius* . . . and by gadgets that actually throw sounds, transmitting them from the location of the speaker to a distant place. . . . They would thus declare that this was a wonderful miracle, but today we know that all of this happened by natural means."[6]

The implications of Moses's alleged ruse in staging the Sinai spectacle are clear to Hurwitz: "Thus all the Torah, and commandments, and laws and ordinances are nothing more than his inventions based on his vast knowledge of the norms and governance of the state. He acted intelligently and with wisdom on behalf of the community. And all of the miracles, all the redemption, wonders, deliverances, and battles that God accomplished for our forefathers in those days and in that time, were engendered by tricks of human wisdom which he [Moses] created by instruments that only he knew how to make."[7]

While Hurwitz's general description of Moses as a natural magician and skillful politician seems conventional enough against the background of deistic and atheistic assaults on biblical miracles, the specific sources of his account are not readily identifiable. Hurwitz hardly mentions the Egyptian roots of Moses's sorcery and legislation as highlighted in the writings of such eighteenth-century authors as John Toland, William Warburton, Karl Reinhold, and Friedrich Schiller. Nor does he seem to be cognizant of recent Christian cabbalistic and Paracelsian renderings of Moses's miraculous abilities. While for Reinhold, for example, the Sinai revelation constituted a sort of open-air enactment of a mysterious Egyptian initiation ceremony, for Hurwitz it was more a technical manipulation of principles of sound and sight as would be known to those adept in the art of pyrotechnics.[8]

Hurwitz's description is perhaps more reminiscent of the infamous treatise of the radical deists known as *Traité sur les trois imposteurs* with its long-repeated accusations that not only Moses, but also Jesus and Mohammed, were insincere deceivers of their respective religious communities.[9] But nowhere is there direct evidence that Hurwitz ever laid eyes on that text. The *Traité* focuses on Moses's ruse in creating the appearance of a cloud and column of fire protecting the people to illustrate his skill as a natural magician. Hurwitz makes a similar argument but instead dwells on the Sinaitic sound and light show supposedly concocted by Moses.

What was critical to Hurwitz in alleging that Moses was a deceiver was the enormous danger this view held for the very foundations of Judaism, the divine origins of the Bible, and the revelation in Sinai. In singling out Moses's supposed abilities as a philosopher knowledgeable in the secrets of nature, he was distinguishing his own understanding of "pious" science from that of the deists and atheists. For Hurwitz, knowledge acquired of the natural world enhances one's appreciation of the mysterious, miraculous, and unfathomable dimensions of God's creation. The scientist can never know enough to understand fully God's ways; thus Moses was acting on God's command, never on his own human initiative.

Hurwitz is so emotional about the consequences of this pernicious understanding of Moses that he even refuses to address co-religionists who hold such a view. He considers them "dead" and beyond hope of ideological or religious rehabilitation. Instead he addresses those who might be confused by such fallacious and destructive views that undermine the miraculous foundations of traditional faith. And he addresses with great respect and reverence Jewish philosophers of past eras such as Baḥya Ibn Pakuda, who assumed it was possible to know God through human investigation. To them he declares, "God desires that our belief in Him be based on the legacy of our ancestors, generation after generation, in a continuing tradition back to the one who actually stood at Sinai."[10] It is this tradition that affirms the veracity of divine miracles and that is supported as well by the testimonies of contemporary non-Jewish scholars.

PHILOSOPHY AS ALIEN TO THE JEWISH TRADITION

Hurwitz's disapproval of the claim that Judaism and metaphysics are compatible, on the one hand, and that God and his miracles can be fathomed through human investigation, on the other, is buttressed by two primary arguments, the first historical and the second philosophical. The first he presents soon after his remarks on Moses and miracles. It is the claim that philosophy was never a part of Judaism in the first place and is alien to its teachings:

> I have heard the view articulated by some Jews that philosophy originated among the children of Israel, from whom the nations of the world learned. But now we learn from them and so return it to its original place. But this is falsehood because the non-Jewish scholars invented it and it is theirs. The ancients were the first who were followed by the scholars of Egypt. There existed a special academy to study philosophy in Egypt, where the great philosopher Solon studied. Afterward, philosophy was transmitted to the Greek scholars of Macedonia and Athens. In the latter place, the great scholars Socrates and Plato were located. Aristotle studied for twenty years in the academy of Plato. The philosopher Anaxagoras at the age of twenty established a great academy in Athens, and there he taught the philosophic methods of Thales. There was also another academy there of the philosopher Antisthenes, and one of his students was the famous Diogenes. The scholars of that city are called [in the language of the rabbis] the scholars of Athens. Afterward, with the destruction of their kingdom and government, all of these academies were destroyed. It [philosophy] was then transmitted to the Arabs and to the Christians of Europe. So from then until now it was in their hands, and the few individuals among the Jews in previous generations who know this wisdom learned it from them.[11]

Leaving aside Hurwitz's sources in constructing this short history of the philosophical tradition,[12] we can acknowledge his competent attempt both to offer a historical synopsis and to challenge a deeply embedded myth among generations of Jewish thinkers that all gentile wisdom originated among the ancient Hebrews. Hurwitz continues:

> Those Jews who declare that the beginning of wisdom originated among the children of Israel thought that this would bring honor and dignity to it [Judaism]. So they put out their arms to wrap themselves in a stolen prayer shawl [*lehitatef be-tallit gezulah*] thinking they would attain a universal name, and with the surplus of honor the Jews then declared that the source of the water came from us. But this is neither true nor honest, since the honor has always been theirs [i.e., that of the non-Jews]. But for any Jew who pursues this honor, it will devour him like the grub so that he will acquire honor like the shade of a rock. For God never gave us this at all; on the contrary, he warned us not to involve ourselves with it lest we be harmed . . . although He gave us a Torah of truth in which everything is included.[13]

Hurwitz's bold declaration that Jews did not invent the philosophical tradition and that it originated in Greece and not Israel is an important statement in modern Jewish thought. For one thing, it displays Hurwitz's commitment to historical accuracy and to telling the "truth" about the past. He provides his readers with a generally correct trajectory of the transmission of knowledge from the ancient Near East to Greece to the Moslem East and the Christian West. For another, it signals a radical break from a central motif of medieval and early modern Jewish thinkers, who required the motif to justify their constant immersion in the study of philosophy and science.

Thanks to the careful study of Abraham Melamed of this grand theme of Jewish self-reflection, we can contextualize Hurwitz's statement as a significant rejection of the notion that all knowledge originated among the Jews.[14] While Melamed points to individuals in Hurwitz's time and earlier who were distancing themselves from this myth of Jewish superiority, he also points out that the myth still had strong adherents well into the nineteenth century. At almost the same time Hurwitz composed his lines, several Maskilim openly employed the myth to justify their scientific and philosophical pursuits and to defend against the claim that Jewish culture was inferior to that of modern Europe. The myth reverberated in Hebrew and Yiddish writing, in pedagogic works, and in books for children.

Particularly noteworthy is the persistence of the myth in the writing of

Isaac Satanov (1732–1804), especially in his *Sefer ha-Middot* (Book of Attributes), published in 1784. In this work on Jewish ethics, Satanov still insisted that the sciences were invented by the rabbis of old, who had even anticipated new fields of knowledge. Moses, so he claimed, invented writing, and a rabbinic sage anticipated Gutenberg's discovery of print by hundreds of years. Satanov, in this regard, echoed a theme articulated by the medieval Judah ha-Levi in his *Sefer ha-Kuzari* (Book of the Khazars), which Satanov not only republished but for which he also wrote an original commentary. Although in his later, more radical writings, he appears to have abandoned this theme, his earlier embrace of the myth is interesting in light of Hurwitz's great antipathy for Satanov—a subject to which we shall return shortly.[15]

Satanov was not alone in expounding the theme of the Jewish origins of philosophy. One finds it also in the writing of Naphtali Herz Wessely, Isaac Beer Levinsohn, Ẓevi Hirsh Chayes, Judah Leib Gordon, and other eighteenth- and nineteenth-century writers.[16] In the opposing camp with Hurwitz were several important thinkers who wrote after the appearance of *Sefer ha-Brit,* such as Peretz Smolenskin and Samuel David Luzzatto. In the case of Luzzatto, the distinction between Hebraism and Atticism became a central thread of his understanding of Jewish theology. We will discuss the parallel postures of Hurwitz and Luzzatto with respect to moral cosmopolitanism in the next chapter. The similarity of their critique of the myth of the Jewish aggrandizement of all knowledge suggests that Luzzatto may have been familiar with Hurwitz's tome.[17] Additional critics of the myth include Aaron Wolfson, Issachar Baer Schlesinger, Judah Leib Mieses, Isaac Samuel Reggio, and others. Bracketing Luzzatto as a traditionalist with an original theology emerging in part from his Italian background, all others were more or less in the maskilic camp, that is, they criticized the myth of Jewish origins as part of their own critique of traditional Jewish values. In this regard, Hurwitz, the alleged "orthodox" Jew, seems to have adopted a radical position to defend the tradition he enthusiastically upheld by diminishing significantly its hegemonic claims as being the source of all wisdom and by acknowledging a substantial body of knowledge fully independent from that of biblical and rabbinic Judaism.

It is important to point out, however, as Melamed emphasizes, that Hurwitz was not completely consistent in his demolition of the myth of Jewish origins. In the passage quoted above, he continues to speak about a Torah that possesses all truth. In the previous chapter, we saw his still-persistent need to show that the rabbis had anticipated several contemporary scientific discoveries. Nevertheless, Hurwitz's contribution to the dismantling of the Jewish claim on philosophy should be acknowledged. His critical historicist view served him well in defending a traditional Jewish faith, but it also excised a significant rationale that had served the apologetic and psychological needs of Jews for centuries. Jews had no exclusive dominion over all learning; they had to share it with the Greeks. Ironically, Hurwitz's attempt to conserve "authentic" Judaism constituted a radically modern revision of the tradition he sought to uphold.

THE CRITIQUE OF BERLIN AND SATANOV

Before presenting his second major argument against the study of philosophy, Hurwitz enlists the testimonies of both earlier Jewish thinkers such as Judah ha-Levi and Joseph Yaveẓ and more recent kabbalists such as Meir Ibn Gabbai, Joseph Ergas, and Emanuel Ḥai Riki.[18] His reference to Riki is not surprising; as we have seen, Hurwitz devoted considerable energy to explicating his writing in a commentary that was published in Cracow after his death.[19] He also returns to his previous reservations about Baḥya's endorsement of philosophy, enlarging the discussion to include other medieval philosophers such as Ibn Ezra, Gersonides, and Maimonides. Although he seems uncomfortable attacking such significant exponents of Judaism, he defends his critique on two grounds. On the one hand, because the truths of the kabbalah had not been widely publicized in their day, the medievals had not been aware of the proper alternative to philosophical study. On the other hand, students of philosophy in his own day had so undermined traditional Jewish faith that defection and atheism were becoming increasingly common. In challenging the study of philosophy among Jews of all periods, he hoped to speak out against the radical exponents of rationalism in his own environment.[20]

At this point, he explicitly targets two individuals and their writings for special disparagement and censure:

> For I have seen in these generations that the faithful are no more [Psalm 12:2] and the deficient methods of philosophy have been strengthened by those who even transgress the principles of Torah and forcefully disseminate their words in public, sinning and causing others to sin. They send their pamphlets to Jewish youth praising and extolling the ways of human investigation and the course of human intelligence. Their intention is to

> trap souls and to win over hearts to atheism [*apikorsut*]. They rejoice in winning over a convert to their cause. They also publish their writings in print, sometimes openly and other times hinting, writing at the end that "the intelligent person will understand" [*ha-maskil yavin*], in order to entice people to do the evil implicit in their allusion. Sometimes they drape their words on a great tree and they write what they want in the name of an ancient scholar who would never have imagined such thoughts. Such is the case with *Sefer Asaf* written by Isaac Satanov. Sometimes they compose their words in the name of some ancient authority and then they write a commentary on his words. Truly ghosts are dwelling there within. The ancient himself exists both internally and externally among the community of ghosts, such as in the case of the book *Besamim Rosh* written by the author of *Kasa Deharsana* [a pie of fish-hash and flour, as in BT Bava Batra 144a] and not Rabbi Asher Ben Yeḥiel. The elders of Israel, teachers of Torah, raise their heads in fear—not the fear of Heaven, but as a supplication that they will not fall from their greatness and be deprived a living, hiding their face in shame without a way to prevent this.[21]

It is this situation that particularly inflamed Hurwitz's anger: "My heart was like a raging fire shut up in my bones. I could not hold it in, I was helpless [Jeremiah 20:9] to restrain my words . . . and thus I declared with this book that it is time to act for God and a time to speak about all these matters discussed until now. . . . Not that I write these words and hope to return those atheists of this generation . . . but I write this all to each Jew who has still not walked in the counsel of evil persons nor stood in the path of sinners."[22]

Hurwitz returns one final time to castigate Isaac Satanov and his work: "However, in this generation it is proper to distance oneself from several Jewish books, among them those of the ancients such as Gersonides and those like him, and among the moderns the books of Isaac Satanov and all that belongs to him. Especially *Sefer Asaf* [should be ignored] because he wrote it to deceive God's people that they might think the ancient Asaph mentioned in the book of Psalms was the author. But this is not the case, since he [Satanov] alone was the author. . . . One who protects his soul should distance himself from all the books he wrote as well as those of his associates."[23]

I have quoted at length here, since I consider this discussion central to understanding the precise cause of Hurwitz's agitation. He speaks not in the abstract about philosophers, atheists, and nonbelievers of his generation, but rather singles out two authors and their alleged forgeries: Saul Berlin (1740–94) and his work *Besamim Rosh*, first published in Berlin in 1793; and Isaac Satanov and his four-volume collection *Mishlei Asaf* and *Zemirot Asaf*, published between 1789 and 1802. Berlin claimed that he was publishing an authentic manuscript of Asher ben Yeḥiel (the Rosh, 1250?–1327), allegedly discovered by a certain Isaac de Molina. Satanov's works represented collections of proverbs which he attributed to the biblical Asaph son of Berechiah, written in the style of the books of Proverbs and Psalms.

Berlin and Satanov were highly educated Maskilim who used their vast erudition in biblical and rabbinic literature to place their enlightened ideas in the mouths of ancients to gain credibility and authority. Modern scholars disagree on how these works should be construed and whether the authors' intentions were cynical and dishonest. For Shmuel Feiner, the public debates that erupted over the credibility of Satanov's and Berlin's writings were the result of the rabbinic elite's fear that its authority was being challenged. Berlin's collection was considered subversive because it showed how the same methods the rabbis employed could be used to reach decisions that were contrary to accepted practice. Some of the responsa even depicted rabbinic culture as "ludicrous, intolerant, and antithetical to enlightenment."[24] Talya Fishman, however, notes the "double-reception history" of Berlin's collection, which some have considered a forgery and some considered a legitimate rabbinic work. Berlin, in her estimation, was a sincere reformer who loved rabbinic culture and was anything but a mere forger.[25] Similarly, Shmuel Werses, in his sensitive portrait of Satanov, does not view the fact that the author hid his identity as insincerity. On the contrary, he claims, Satanov's intention was honorable, in that he viewed his designation of Asaph as the author of his collection as a legitimate Jewish practice.[26] In neither case can these writings be described simply as outright assaults on rabbinic Judaism. Rather, they constitute a complex dialectic position of simultaneous love and hate for the tradition in which these authors wrote.[27]

Whatever the intentions of Berlin and Satanov, Hurwitz concluded that they were insincere forgers who had manipulated the sacred tradition of Judaism and its literary legacy for their own nefarious purposes. Their sin of holding enlightened views was compounded by their misrepresentation of what the Jewish tradition in fact says. Instead of openly and transparently challenging the tradition in their own voices, they pretended that it was the tradition itself that spoke. This fraudulence and deception infuriated Hurwitz. On the battlefront between traditionalists and modernists, one needs to play by the rules of the game, preserving the boundaries between the two warring camps. A critic of the tradition should not masquerade as an upholder of the tradition and vice versa.

Let us recall Satanov's appropriation of the myth of the Jewish origins of philosophy, discussed above. Hurwitz's critique of Satanov's and Berlin's alleged forgeries was fully consistent with his general objection to this myth. In both cases, they misrepresented and falsified the integrity of Judaism. Philosophy, in his view, was an alien import that needed to be amputated from Judaism. When it was represented as part of the tradition, either overtly or through dissimulation and deception, its unnatural status needed to be identified, then purged. In his effort to protect and preserve what he understood to be a monolithic tradition, Hurwitz would take advantage of all the literary and historical tools at his disposal to expose unworthy distortions of its ancient texts and contexts.

KANT AND THE LIMITS OF METAPHYSICAL THINKING

Hurwitz's second argument against the place of metaphysics in Judaism is based on his personal discovery of the philosophy of Immanuel Kant, specifically his critique of metaphysics as laid out in *The Critique of Pure Reason*, published in 1781. Hurwitz, however, never studied the work himself.[28] Instead he relies and expands on Solomon Maimon's short summary of Kant in the introduction to his *Givat ha-Moreh* (The Hill of the Guide), first published in 1791. But despite the brevity of his primary source, Hurwitz well understood the thrust of Kant's general argument, summarized it reasonably well, and utilized it in his own assault on the place of philosophy in Judaism. Here is how he introduces the subject:

> It is known throughout the entire land that in 1781 a very wise man with great understanding arose among the gentile philosophers in the city of Königsberg in the kingdom of Preussen [Prussia]. [He was] diligent, clear-headed, and acute and his name was Kant. He wrote an outstanding and very deep book on science, and the book became very well known in all lands. He wrote in the beginning of his book that he addresses the generations of philosophers from Aristotle on, those who preceded him, and the most recent ones of this generation. He challenged them to tell him what

> was new in their investigations and in their theoretical proofs regarding what is beyond nature, called metaphysics. He asked that they present him an indubitable proof of an absolute truth which all can accept so that he can propose a proof that contradicts it, since all those who devise speculative proofs are mistaken. If this is the case, what do philosophers with all their theoretical wisdom accomplish with demonstrations and proofs? Thus he described in his book the works of the great philosopher [Christian] Wolff [1679–1754] and the famous philosopher [Gottfried Wilhelm] Leibniz [1646–1716], who are the best known among contemporary philosophers: they build edifices like the toys innocent children construct on the table from paper or cards called *karten*, which the wind of the human mouth blows away [cf. Psalm 1:4].[29]

Hurwitz proceeds to summarize Kant's philosophy, singling out the four antimonies, the seemingly contradictory metaphysical propositions that human beings can never resolve with absolute certainty: the limitation of the universe with respect to space and time; the theory that the whole consists of indivisible atoms; the problem of free will in relation to universal causality; and the existence of a necessary being. Hurwitz describes them with the terminology of Jewish philosophy: the existence of a necessary being is called "God's existence"; the limitation of the universe refers to the "Creation of the world"; and so on. Following this concise synopsis, Hurwitz, relying on Maimon's summary, underscores Kant's major argument of the impossibility of human beings acquiring any new idea that is beyond experience: "For we lack the tools and means and instruments to investigate and to search to know a thing that is not based on the senses and that is beyond nature." He is thus thoroughly convinced "that because of the words of this gentile philosopher, the bastions of all philosophy have been destroyed [cf. Jeremiah 50:15] and her foundations have fallen and will never rise again."[30]

At this point, however, Hurwitz introduces a dissonant note that appears to challenge Kant's seeming victory: "I lifted up my eyes to see that the philosopher [Solomon Maimon], the author of *Givat ha-Moreh*, his commentary on the *Moreh Nevukhim* [of Moses Maimonides], in his introduction, had attempted to defend philosophy and to save it from the approach of Kant and to rescue it from this prison."[31] What follows is the almost complete quotation of Maimon's summary of his own position, which in *Givat ha-Moreh* immediately follows his short descriptions of Leibniz and Kant in the introduction. Maimon claims, in deviating from Kant, that our philosophical assumptions are based not on sensibility itself but on what he calls the foundations of sensibility, that is, tiny entities called monads, an intellectual construct conceived by Leibniz. These ultimate elements of the universe are not related to place and time and are indivisible. Although human intelligence cannot relate to the senses themselves, it grasps the value and relation between the foundations of the senses. By this method Maimon sought to preserve the relation between the forms of cognition and the senses, which are not objects of the intellect.[32]

Hurwitz was clearly not equipped, on the basis of this short passage alone, to comprehend the subtle and complex criticism that Maimon offered to Kant's critique of metaphysics. But he was not about to accept Maimon's assault on Kant based on Leibniz's monadology: "And I the writer, after studying his words, declare that there is nothing of substance in his responses and his words have no value and are vanities, while the words of Kant are firm as a mirror of cast metal [Job 37:18]. This is because he [Maimon] relied on Leibniz's notion of tiny entities, and in reality these entities never existed. Leibniz himself concocted them and declared to his students: 'Know and believe me that there exist in reality simple individual entities called *einfache Wesen* or *Monaden*.'" After describing their characteristics according to Leibniz—again based exclusively on Maimon's summary—he declares:

> And now, enlightened reader and understanding man, tell me, who can believe in this report and who actually saw these things? . . . How did Leibniz actually know this? What certain proof did he possess? Did he have a tradition from his forefathers in his hand, or perhaps he learned it in a dream or a special spirit possessed him? . . . And if you claim that many famous scholars already held his views and followed in his footsteps, I will answer you by asking: "Was there any greater scholar than the late Maimonides, who upheld the views of Aristotle and rested on them as on a stone pillar?" And not only him, but many scholars among the nations who came after him, generation after generation, all believed in his words . . . to the extent that anyone who rejected them and denied their assumptions was mocked and ridiculed in their eyes as one who denies what is verified by the senses. Nevertheless, recent philosophers arose among the nations who contradicted his assumptions and negated his views because they subjected his words and assumptions to the furnace of investigation, and they did not hold up to the test of experience.

Hurwitz's final objections evoke common sense and history:

> And didn't Kant know that Leibniz had invented and created the monads, since he lived after Leibniz, read all his books, and investigated all his words? . . . He simply consciously distanced himself from this atrocious idea, actually imagined and crazy words having no relation to reality other than in the mind of Leibniz himself. And one more point I ask of the author of *Givat ha-Moreh*: What about all the generations who preceded Leibniz, the father of monads? . . . And what were the objects of philosophy for all of them, the ancient and modern philosophers until close to Leibniz's own era who lacked the monads, since they were only now created, not earlier? For it was he [Leibniz] who commanded, and they were created and they were upheld by his followers as an eternal law![33]

Hurwitz, as we have said, was no philosopher. Nevertheless, his arguments appealing to Kant are intelligent, accurate, and creative. Most important, he understood precisely the import of Kant's critique of metaphysics in defending the veracity of the Jewish revelation as interpreted by the kabbalistic tradition. Hurwitz understood it just as Kant had articulated it. In the preface to the second edition of the *The Critique of Pure Reason*, published in 1787 (a text that Hurwitz likewise probably never saw), Kant wrote:

> [From what has already been said, it is evident that] even the assumption—as made on behalf of the necessary practical employment of my reason—of God, freedom, and immortality is not permissible unless at the same time speculative reason be deprived of its pretensions to transcendent insight. For in order to arrive at such insight it must make use of principles which, in fact, extend only to objects of possible experience, and which, if also applied to what cannot be an object of experience, always really change this into an appearance, thus rendering impossible all practical extension of pure reason. I have therefore found it necessary to deny knowledge, in order to make room for faith. The dogmatism of metaphysics, that is, the preconception that it is possible to make headway in metaphysics without a previous criticism of pure reason, is the source of all that unbelief, always very dogmatic, which wars against morality.[34]

Hurwitz, like Kant, was denying the human pretension to knowledge in order to make room for faith. And in this embrace of Kant, he was hardly alone among his co-religionists. As Paul Franks notes, no non-Jewish philosopher has been more central to Jewish philosophy than Kant. For one thing, Kant's career coincided with the entry of Jews into German academic and intellectual life. For another, a general affinity existed between Jewish philosophy and Kant due to the long involvement of Jews with the Platonic tradition. Finally, certain distinctive features of this Kantian Platonism were especially susceptible to a distinctly Jewish interpretation, such as the law, messianism, and the notion of the divine unity.[35]

David Ellenson has explored the uses of Kant among German Orthodox Jewish thinkers at the beginning of the twentieth century, including Joseph Wohlgemuth, Nehemiah Anton Nobel, and Isaac Breuer. This appreciation of Kant reached its zenith in the writings of Breuer, who, like Hurwitz, studied the kabbalah. He wrote that "God caused to rise among the nations the exceptional man Kant, who, on the basis of the Socratic and Cartesian skepticism brought about that 'Copernican Turn' whereby the whole of man's reasoning was set in steel limits within which alone perception is legitimized. Blessed be God, who in His wisdom created Kant! Every real Jew who seriously and honestly studies *The Critique of Pure Reason* is bound to pronounce his 'Amen' on it."[36]

Allan Mittleman's account of Breuer's relation to philosophy succinctly captures Hurwitz's posture as well:

> First, Breuer wants to overthrow the modern preeminence of understanding [*Verstand*] and reason [*Vernunft*] as arbiters of religious meaning. He wants to take Judaism outside of the bounds of reason alone, not by a leap of faith or a plunge into mystical absorption, but by so delimiting the bounds of reason that very little human meaning can subsist therein. There is a postmodern impulse at work in his epistemology. He wants to dismantle religion within the limits of reason alone. He wants to destroy the contemporary Jewish mythos of Judaism as a religion of reason. The alternative is not a religion of unreason, but a religion that transcends reason."[37]

Pinḥas Hurwitz, a Jewish writer from Vilna and Cracow, had anticipated by more than a hundred years Breuer's Jewish theology both with respect to the place of kabbalah in modern Judaism and in his appropriation of Kantian philosophy. One wonders whether Breuer himself profited from a reading of *Sefer ha-Brit.*

Despite his limited exposure to the philosophy of Solomon Maimon, Hurwitz captured accurately his deviation from Kant, at least in part. Maimon indeed was a dogmatic rationalist insofar as he endorsed the standards of reason found in Leibniz and Spinoza, but he also remained empirically skeptical insofar as he agreed with David Hume that such standards of reason are never met in our experience of the world. Hurwitz was clearly unaware of this skeptical bent. But Maimon, like his medieval namesake Maimonides, was committed to the world of the mind and believed in the notion of intellectual progress. For him, it made no sense to follow Kant's claim by denying knowledge in order to make room for faith. Rather, Maimon rehabilitated the Aristotelian notion that the highest virtue and pleasure are found in philosophical contemplation.[38]

While deficient in modern philosophy, Hurwitz knew his Maimonides well and sensed in Maimon the medieval philosopher's ideal of intellectual perfection being the ultimate human end. For Maimon, following his teacher, a human being becomes closer to and more like God when he develops his intellectual capacities. And Hurwitz, not only in the critique of Maimon quoted above but elsewhere in *Sefer ha-Brit* as well, upbraided his philosophic contemporary for mistaking intellectual perfection for the ultimate redemption of the human soul.[39]

Despite his reverence for the sage of Fustat, Hurwitz lamented the powerful hold Maimonideanism held over Jewish thinking for centuries. But intellectual fashions come and go, and thus ultimately Aristotle and Maimonides were overturned by Leibniz and Wolff, who were themselves undermined by Kant. The study of history not only revealed the finitude of human systems of rationality; it also exposed the foreign and estranged nature of the philosophical tradition in Judaism in general. Through Kant and through a rigorous historical probing of the Jewish intellectual past, Hurwitz had constructed a powerful rational argument about the limits of reason and philosophy in the Jewish tradition.

HURWITZ ON FAITH

We cited Shmuel Feiner at the beginning of this chapter, who claimed that in the contest between science and faith that Hurwitz allegedly fought, he always chose faith. Before closing this chapter, I would like to briefly look at this notion of faith as Hurwitz discusses it and try to understand what he meant by it in the context of his "contest" with philosophy and science.

Following his discussion of Kant, Hurwitz offers the following definition:

> Know, my brother, that the definition of the word "faith" [*emunah*] and its interpretation is the following: It is what a man knows and accepts in his heart and pictures in his imagination as actually existing in reality outside of what he thinks and what he feels in his heart. It is not something which a human can know by rational proof nor can he experience it from one of his senses, because what is known by proof or by sensation is knowledge, not faith. So, for example, a man would not say, "I believe that this table is made of wood." Rather, he would say, "I know that this table is made of wood." It [faith] is something a person can know by way of a tradition received from his forefathers or his teachers who related that there was such a specific thing in reality. And the heart then accepts it, to believe it honestly and uprightly.[40]

I have not identified the precise source of this simple definition, but its general context is perfectly clear. One finds discussions of this notion in medieval Jewish philosophy beginning with Saadia. Closer to Hurwitz's world, one might look to the definition of faith in John Locke's writings or reverberations of his position in the writings of such Jewish contemporaries of Hurwitz as Mordechai Gumpel Schnaber Levison.[41] Following Hurwitz's long discussion of the sources and limits of human knowledge based on Kant, he presents a definition of faith that stands in sharp contrast to metaphysical thinking and sensory perception. In other words, Hurwitz's notion of faith is just as culturally conditioned as are his notions of the senses and of metaphysics. He had appropriated the discourse of his philosophical contemporaries, and it was against this epistemological background that he articulated his notion of Jewish faith. His understanding of faith, in this regard, was not at loggerheads with science. Rather, it was an essential part of the scientific vocabulary of an educated person and could only be understood in its dialectical relationship with philosophy and sensory experience. The term *faith* and its precise definition were just as alien to the imagined tradition Hurwitz attempted to defend as philosophy and science were.

I hope it is evident to the reader that Hurwitz's articulation of Judaism was intellectually rigorous, informed by the latest historical and philosophical knowledge of his day, and constituted a creative response to the cultural and intellectual challenges of his environment. Labeling it an "orthodox" project hardly captures its complexity and richness. On the contrary, Hurwitz genuinely accepted aspects of the larger cultural world in which he lived, such as history and science, while rejecting others, such as a certain kind of philosophy, namely metaphysics. He was critical of the latter because of its pretension to know everything, its refusal of epistemological limits, and its claim of self-sufficiency. At the same time, his discourse in the name of Judaism was informed by philosophical and historical thinking, an unmistakable testimony to his avid participation in an animated and learned conversation with the modern intellectual world in which he lived.

CHAPTER 5

The Moral Cosmopolitanism of Pinḥas Hurwitz

SOME INITIAL CONJECTURES

THE most unusual and original chapter of Pinḥas Hurwitz's *Sefer ha-Brit* is undoubtedly the one he calls "Ahavat Re'im" (The Love of Neighbors, based on Leviticus 19:18), the lengthy thirteenth chapter of book two. In one instance, it was even printed as a separate publication, underscoring its significance as a moral treatise in its own right. In 1842–43, Sa'adi ha-Levi, the renowned Salonika publisher of more than two hundred books between 1840 and 1902 and the founder of the journal *La Epoca*, published the chapter in a Ladino translation as part of an anthology of three works with the Hebrew title *Sefer Darkei ha-Adam* (The Book of the Ways of Man), surely underscoring its universal appeal.[1]

More recently, Resianne Fontaine has presented an accurate and detailed synopsis of the chapter along with a reflection on its context and the author's motivation in writing it.[2] It is certainly far from clear why Hurwitz wrote such a chapter in the first place, why he included it in this large compendium of naturalistic-mystical reflections, or why he placed it almost at the end of his book, at the very point where he was about to reveal to his readers his final instructions for imbibing the Holy Spirit and attaining the level of prophecy. Hurwitz himself was fully aware of the novelty of his chapter and noted that he had great difficulty finding sources that discussed his position, either in Jewish or non-Jewish texts:

> I was determined to search for this exalted matter [the love of neighbors] among ancient and modern Jewish authors, and I examined both small

and big books, new and old. I looked and didn't find except for the general reference of two or three words, as it is written in the Torah in the three words: "And you shall love your neighbor as yourself." I also searched among non-Jewish authors, but I also did not find anything except for some direct thoughts of Cicero. I was thus motivated to declare that I will discuss this matter in detail in this book, explaining its meaning, all its aspects, considerations, and definitions, and I will elaborate a bit more on this and devote an entire discourse to it in order to awaken the hearts of my fellow Jews to uphold the bond of community and society of all humanity and to alert them to this great responsibility, which is weighty, and to instruct my people in the ways of brotherhood and friendship so that then they will become wise and enlightened. Then the eyes of the blind will be opened wide.[3]

There is no reason to discount Hurwitz's claim that what he wrote was original and had not been drawn from any previous author. But the historian is still left with the challenge of explaining what motivated the author to articulate so independent a position and why it was crucial to him to locate it so prominently in his vast educational and moral project in the first place. Since Hurwitz remained relatively silent about these questions, one is left only with a few vague hints to attempt to contextualize his message. With these meager clues in mind, I offer here my tentative reflections on the genesis of this striking chapter.

HURWITZ'S MESSAGE IN THE CONTEXT OF PREVIOUS AND CONTEMPORANEOUS JEWISH ATTITUDES TOWARD THE CHRISTIAN OTHER

Drawing in part from the excellent summary of Resianne Fontaine, I begin with a brief reiteration of Hurwitz's general position. He opens with a definition of neighborly love as that of loving the entire human race regardless of which nation or language one belongs to,

since every human being is made in the divine likeness and image and occupies himself with the betterment of the world by building, plowing, planting, trading, selling, or in a craft or in conceiving of insights and strategies by which he acquires worldly needs. He generates thoughts so as not to neglect the needs of other creatures, and he prepares the world in his wisdom, envisioning and investigating and replicating wonderful instruments and creations through his own ingenuity by which he labors. For through these things, the world stands properly and exists in perfection, while all these things exist because God created and made them and they are very good to every human being.[4]

The harmonic interactions of human society include all kinds of people, with two notable exceptions: wild savages and criminals.[5] For all others, societies rest on their practice of mutual love and respect, comparable to the harmonious relations of limbs and organs of a living creature.[6]

Hurwitz's notion of the obligation to love one's neighbor is based not only on rational calculations, so he claims, but on the insights of the Bible and Jewish tradition as well. With respect to the Bible, he attempts to explain away its elimination of the seven Canaanite nations as an exception to the general rule, remaining emphatic that the Leviticus injunction does not mean only Jews but all human beings (although he does acknowledge the paucity of Jewish sources supporting this position).[7]

Having laid out his theoretical position, he then offers a detailed discussion of the practice of human relations—familial, social, economic, and political. Here he explains both how human beings can avoid harming each other and how, in a more active sense, they can enhance one another's lives.[8] He also explores the limitations, challenges, and priorities in living a practical and prudent moral life.[9] In these elaborate discussions, one senses a subtle shift in emphasis from humanity in general to the Jewish community in particular. This is especially evident in his treatment of senseless hatred, where he mentions explicitly the tensions between ashkenazic and sephardic Jews and the need for unity within the community.[10] Hurwitz closes this lengthy discussion by returning to a theme raised in other parts of his book: the idea that scholars of Torah are not exempt from moral accountability and that wisdom or erudition alone is no substitute for ethical sensitivity and responsibility.[11]

Hurwitz's claim that the position he was championing could be located within traditional Jewish sources, especially that of his most revered sage, Ḥayyim Vital, was, however, not so simple to demonstrate. In the case of Vital, although Hurwitz cites him correctly—"One should love all creatures, even the non-Jew"[12]—the insertion of the word *even* implies that the moral status of the non-Jew is not equal to that of the Jew. Hurwitz's position is surely a stronger, more inclusive formulation of the notion of universal human love than that of Vital in *Sha'arei Kedushah*. The few other sources he cites to substantiate his position within the Jewish tradition were likewise weaker expressions of his position.[13]

By examining more systematically the long history of Jewish attitudes toward the non-Jew, especially in the context of Jewish exegesis on Leviticus 19, one comes to appreciate the uniqueness of Hurwitz's bold articulation even more. According to Ernst Simon, while the Bible is not decisive on the matter, Jewish commentators, from Maimonides, Nachmanides, and other medievals all the way up to modern scholars, almost always interpret *neighbor* as Jew. Simon's apparent dissatisfaction with this traditional position led him to ask whether Jews require a higher ethical stance beyond the traditional one.[14]

In Elliot Wolfson's *Venturing Beyond,* a masterful summary of premodern Jewish attitudes toward the non-Jew, especially as found in kabbalistic sources, the same picture emerges but in even more stark and negative terms. In kabbalistic ethical writings, the term *adam* (human being) always refers to a Jew, the fate of the gentile being a matter of indifference. Wolfson speaks of the demonization of the gentile in these sources, reflecting not just narrow chauvinism but even xenophobia. In *Sefer ha-Zohar,* for example, the Jew appears as a human being in contrast to the beastly character of gentiles. So pervasive was Jewish ethnocentrism than even kabbalists influenced by Maimonides, who identified the essence of man as consisting in the faculty of reason, understood this faculty to apply specifically to Jews. Wolfson presents a large inventory of kabbalists and other Jewish writers who fully preserve the distinction between pure Jews and impure gentiles, including Abraham Abulafia, Moses Cordovero, the Maharal, Isaiah Horowitz, Menasseh Ben Israel, Saul Levi Morteira, and Ḥayyim of Volozhyn. It is only the Christian kabbalists, Wolfson points out, who reinterpret kabbalistic texts in universal terms.[15]

Jacob Katz, in his book *Exclusiveness and Tolerance,* charts the double moral standard of Judaism toward the non-Jew until the late thirteenth century, at which point Menaḥem ha-Meiri injected a decisive exegetical shift, distinguishing between ancient pagans and contemporary monotheists, idol worshipers and Christians.[16] For Katz, and for other scholars in dialogue with his position, Meiri opened a long trajectory of Jewish interpretation disassociating Christianity from idolatry and demonstrating a growing appreciation of the values common to both religions.[17] Katz, for example, singled out such Jewish thinkers as Eliezer Ashkenazi, Moses Rivkes, Yair Bacharach, Jacob Emden, Jonathan Eibeschütz,[18] Ezekiel Landau, Eliezer

Fleckeles,[19] and others from the sixteenth century on who openly adopted a more tolerant view of the religion of the majority among whom they lived. Yet there were limits to these positions, since the emphasis was always on correct behavior and toleration, not on love. The Jewish legal scholars were particularly sensitive to the need for preaching fairness while preserving the proper social and religious boundaries between Jews and Christians. Even the enlightened Moses Mendelssohn spoke about toleration, not love, in extolling the common humanity of the two religious faiths.[20]

By the eighteenth century, the question of the treatment of the "goy" in Jewish literature and law was held up for ridicule by such learned Christian scholars of Judaism as Johann Andreas Eisenmenger and Johann Christopherus Wagenseil. Jews responded apologetically, as they had done for centuries, but now with greater intensity and alarm. Such rabbis as Yeḥiel Michal, Benjamin Woolf, and Jacob Ẓevi Mecklenburg, each in his own way, underscored the monotheistic nature of the Christian faith and argued that the laws separating idolaters from Jews in antiquity no longer applied to contemporary Christians, who were to be counted among the righteous of the world.[21] Jacob Emden also stressed the morality of enlightened Christians in enhancing the world, while Abraham Loewenstamm offered an apologetic tome written in German that attempted to deflate the force of rabbinic dual moral standards.[22] Under the pressure of the Napoleonic Sanhedrin, Aaron of Worms articulated the notion of fraternity in the biblical designation of "neighbor," understanding that citizenship in France had to supersede any ethnic or religious group loyalty.[23]

In other words, by the time Hurwitz composed his chapter on universal love, several of his Jewish contemporaries had already wrestled with the new social demands of living in close proximity to their Christian neighbors and were openly declaring a greater appreciation of Christianity as a religion and recognizing the enlightened behavior of many followers of the Christian faith. Hurwitz was hardly the first to articulate a language of common humanity in discussing intergroup relations between Jews and Christians. But his position still stands out for its bold employment of the language of love in describing these relations and in insisting that ultimate religious fulfillment—that is, imbibing the Holy Spirit—could not be reached without a total commitment to moral cosmopolitanism.

To identify a close analogue to Hurwitz's formulation, one must look beyond his generation into the nineteenth and early twentieth centuries, especially to two well-known Italian Jewish thinkers, Samuel David Luzzatto and Elijah Benamozegh.[24] Luzzatto's moral commitment to non-Jews stemmed from his notion of the Italian *compassione* or the Hebrew *ḥemlah*, a sentiment that all human beings share and that was reinforced in Judaism by the divine revelation at Sinai: "The compassion that Judaism commends is universal. It is extended, like God's, to all of his creatures. No race is excluded from the Law, because all human beings, according to Judaism's teachings, are brothers, and children of the same Father, and are created in the image of God."[25]

Luzzatto was quite aware of the double moral standard between Jews and gentiles that had emerged in classical Judaism and within rabbinic law. And remarkably, he was willing to challenge the rabbis if their positions violated his own standards of universal morality. He explained their moral limitations by the difficult historical circumstances Jews had experienced over the centuries: "Whatever proposition or story that could be found in the Talmud . . . which would be in opposition to the sentiments of universal humanity and justice and which are suggested equally by nature and by Sacred Scripture, must be regarded as neither of the dictates of Religion or of Tradition, but as regrettable insinuations of the calamitous circumstances, and of the public and private vexations and cruelties to which the Jews were subjected in the barbaric sentiments."[26] For Luzzatto, a universalized biblicism based on the ethics of human sentiment allowed him to define a Jewish morality over against the weight of historic rabbinic traditions vis-à-vis the gentile.

Elijah Benamozegh's commitment to a universal morality rested on foundations more metaphysical and kabbalistic than those of Luzzatto. In this regard, he was closer to the outlook of Hurwitz, whose moral cosmopolitanism came directly from his mystical orientation. Benamozegh's principal concern was to demonstrate that a universal vision of Judaism might address the challenges facing human civilization more effectively than Christianity. The universality of Judaism as expressed through the kabbalah represented a synthesis of all the world's religions. True faith in one God not only united all religion but also harmonized science and religion. Judaism was a particular path leading to a unity of God and humanity that included the best manifestation of all human spiritual insights.[27]

Hurwitz would have considered these elements of Benamozegh's vision foreign to his simpler faith in the superiority of Judaism and its unique revelation at Sinai. Nevertheless, Benamozegh's universalism grounded in kabbalah and science at least suggests a linkage back to *Sefer ha-Brit* and to

Hurwitz's confidence in the power of kabbalah to engage both the natural and moral orders. It also situates Hurwitz's perspective in an Italian cultural matrix. One need only recall his fascination with and reliance on early modern Italian Jewish kabbalists, especially Emanuel Ḥai Riki and Joseph Ergas, in constructing his own understanding of Lurianic kabbalah.[28] In a wider sense, an encyclopedia of the sciences informed by kabbalah was primarily a project of Italian Jewry.[29] In that Hurwitz, an askhenazic Jew from Vilna, no doubt drew inspiration from Italian models, it seems relevant to connect his own Jewish vision with those of these two more complex (and diverging) Italian Jewish thinkers of a later generation.

THE IMAGE OF A HEROIC GERMAN PRINCE AND THE MORAL VISION OF *SEFER HA-BRIT*

A comparison of Hurwitz's "Ahavat Re'im" with earlier and later Jewish expressions of universal morality is useful in contextualizing the work within the history of Jewish attitudes toward the Christian "other," but it hardly provides precise clues regarding its specific genesis or the author's sources of inspiration. A search for such clues within the text itself yields little, with the exception of one dramatic narrative that I here quote in full. It is located near the beginning of the chapter and provides a small but significant glimpse into a cultural world apparently close to the author and to some of his Jewish contemporaries:

> Experience teaches us many times than even the sons of kings and the rulers of the earth put themselves in mortal danger, confronting fire and water in order to save a person who fell into a river or in order to rescue from fire when a person's house was burning and other such examples. Such is the case that happened in the city of Frankfurt on the Oder in 1785 in the month of Iyar on the seventeenth. The Oder River was rising higher and higher, flooding the area, overflowing its banks. Several villages were overwhelmed with powerful floods overtaking homes on the shore. Many pieces of wood and windows unhinged from their homes floated through the streets of the city, and several people drowned and lost their lives. And in the midst of them all floated one large piece of wood, and on it stood a living person who called to the men standing on the banks of the river to help and save him. But it was impossible for them to do so because of the swift current. When the great and honored prince duke Leopold, who was

> then the commandant of the city, saw this, he immediately commanded his navy to accompany him in boats to save this man. But they were not able and said to him, "We are incapable of doing this because of the powerful water currents and high waves. No ship can make it through." When he heard this, the duke immediately took a boat, risking his life to save this person. Before reaching halfway across the river, his boat capsized and he was overtaken by the many waves. The boat sank, and the righteous prince was lost and no one could save him.[30]

Hurwitz presented this story within a discussion of whether reason requires human beings to feel a moral responsibility to their society. To this Hurwitz answered affirmatively:

> Every human being has an absolute duty to always promote decency and righteousness within his own society so that the community will survive, and he should bother to insure that evil be distanced from it and that no member will be hurt. He should seek its peace and good welfare all the time in order to maintain the social covenant whose bonds are connected in fulfilling appropriate and fitting deeds to gain benefit for them and from them. And thus it can be explained rationally that this is an absolute obligation for all persons who are members of an eternal covenant.

Hurwitz also argued that this obligation was based on a natural disposition on the part of human beings to do goodness: "If a person examines his soul and natural makeup, he will discover an inclination and a desire to do righteous things in the eyes of everyone to improve and to impact the other." The story of Leopold followed as an illustration of this second point, closing with the words "Thus this obligation is also demonstrated as a natural tendency."[31]

Hurwitz's illustrative story of a person naturally inclined to do good was based on an event that happened on April 27, 1785, the drowning of the young prince Maximilian Julius Leopold von Braunschweig-Lüneburg (b. 1752), who perished in a flood on the Oder River trying to save others who were threatened by the disaster. Leopold was the youngest son of Duke Charles I and his wife, Philippine Charlotte of Prussia, sister of Frederick the Great. He received an enlightened education from well-known tutors and even was accompanied by Gotthold Ephraim Lessing on a specially arranged educational trip to Italy in 1775. Leopold was also an active Freemason and had built a reputation as a humane prince whose primary concern was the welfare of his citizens and the alleviation of human suffering.[32]

His death caused quite a sensation among the literary elite in Germany and beyond.[33] The news of the drowning occasioned an outpouring of sympathy and grief, expressed in poetry, art, commemorative medals, and even full-scale monuments to honor this new embodiment of the *Menschenfreund*, a man who was alleged to have declared right before his death: "I am a man as you are, and this is a case of saving a man's life." His eulogizers included such literati as Goethe, Herder, and Schiller. So Goethe wrote:

> Thou wert forcibly seized by the hoary lord of the river
> Holding thee, even he shares with thee his streaming domain
> Calmly sleepest thou near his urn as it silently trickles
> Till thou to action art roused, waked by the swift-rolling flood
> Kindly be to the people, as when thou still were a mortal
> Perfecting that as a god, which thou didst fail in, as man.[34]

French literati were also moved by Leopold's heroic act. The French Academy designated him a royal hero in 1785, and sermons were preached in his honor from the pulpit of Versailles. When the comte d'Artois offered a prize for the best poem on Leopold, more than 150 French poets submitted their work.[35] Artists also made Leopold a figure larger than life, especially the well-known Daniel Chodowiecki in a series of paintings depicting the prince's actions and ultimate demise. Pierre-Alexandre Wille, too, captured Leopold's last moment in a painting, and Johann Gottfried Jacob Stierle memorialized his death in several gold medallions that he designed.[36]

In short, Hurwitz had found more than a good story about a prince to illustrate his point. He had found a universal icon symbolizing the best of humanity who had instantaneously become an object of reverence for a community of intellectuals smitten by Enlightenment ideals, especially those engendered by Freemasonry. Perhaps it might be worth recalling the larger context in which the famous prince was celebrated. One of his encomiasts, Friedrich Schiller, composed at about the same time the popular "Ode to Joy" for the periodical *Thalia*, later to be immortalized by Beethoven's Ninth

Symphony. Written to Schiller's friends in Leipzig, this poem reflected the rhetoric of both the late German Enlightenment and the Freemasons. To quote only one stanza:

> Be embraced, millions!
> This kiss to the entire world!
> Brothers—above the starry canopy
> A loving father must dwell.
> Whoever has had the great fortune,
> To be a friend's friend,
> Whoever has won the love of a devoted wife,
> Add his to our jubilation!
> Indeed, whoever can call even one soul
> His own on this earth!
> And whoever was never able to must creep
> Tearfully away from this circle.[37]

As Robert Beachy has described, Freemason rhetoric made a deep impression on German court culture in the final decades of the eighteenth century, particularly its focus on moral cosmopolitanism and the improvement of humanity. It peaked prior to the rise of German romanticism and protonationalism during the revolutionary era. By 1800 there were 350 lodges in Germany alone, comprising some 20,000 Masons, including a broad spectrum of educated people attracted to the powerful rhetoric and sociability that Masonry offered.[38] If one notes, for example, Goethe's allusions to the Masonic theme of universal brotherhood, the Masonic traditions evoked in Mozart's *Magic Flute*, or, most prominently, Lessing's Masonic dialogues and his call for universal tolerance in *Nathan the Wise*, there is no doubt that the theme of universal love pervaded the literary and artistic ambiance of that era.[39] And Leopold, a protégé of Lessing and a devotee of Freemasonry in general, captured in his tragic death this acute sensibility of the family of humanity and the need to overcome national and ethnic barriers in order to embrace that universal emotion. Could Hurwitz have known about *Nathan the Wise* or the close relationship between Lessing and Hurwitz's coreligionist Moses Mendelssohn?[40] Was his call for loving the non-Jew a not-so-subtle criticism of Jewish separatism and chauvinism, a response to ideas articulated by Freemasonry and so boldly dramatized by Lessing? Might Hurwitz have learned about Freemasonry from his dear friend Jacob Hart in London, a member of his local Masonic fraternity whose opinions were cited at least four times in *Sefer ha-Brit*?[41] Would he have retold the story of Leopold for his Hebrew readers without appreciating the deeper meaning of the prince's heroic image among his contemporaries? Alas, such questions have no certain responses.

Whatever Hurwitz meant to convey in his daring chapter on loving one's neighbors, it needs to be balanced against comments found elsewhere in *Sefer ha-Brit*, comments that seem to perpetuate strong feelings of Jewish superiority and estrangement rather than universal brotherhood among all peoples.[42] But despite these seeming inconsistencies, his ringing endorsement of universal love seems linked to the broader moral cosmopolitanism championed first and foremost by Lessing.[43] The positioning of the Leopold story within Hurwitz's narrative testifies to the fact that he was aware of a larger moral perspective shared by his contemporaries through which he could enhance a Jewish reading of the biblical text. Whatever his precise source, Hurwitz was telling his reader that a Christian prince steeped in the principles of Freemasonry had much to teach Jews about their own ethical behavior.

Hurwitz was not the only Jew to appreciate Leopold and his exalted moral standing. Soon after Leopold's drowning, a pamphlet was published in Frankfurt an der Oder entitled *Leichenrede des Herzogs Maximilian Julius Leopold von Braunschweig*, the German translation of a Hebrew sermon delivered on May 9, 1785—a little more than two weeks after the prince's death—by one Rabbiner Joseph, rendered by Eli Naphtali Cohn. Rabbiner Joseph was none other than Joseph ben Meir Teomim (c. 1727–92), a distinguished Talmud scholar and since 1781 the rabbi of Frankfurt. Teomim was especially well known for his commentary on the classic code of Jewish law, the *Shulḥan Arukh*, called the *Peri Megadim*. In addition, Teomim published Talmudic novellae; expositions of the codes of Alfasi and Maimonides; works on Talmudic hermeneutics; collections of ethical sayings, sermons, and responsa; and more.[44]

Teomim was neither a philosopher nor a Jewish thinker known for his enlightened tendencies. Rather, he wrote exclusively within the domain of rabbinic law and exegesis. That such a traditionalist would compose a eulogy mourning the tragic death of Frankfurt's famous prince and allow it to be circulated in German translation seems remarkable in its own right. Reading the seemingly genuine outpouring of grief for the departed leader leaves the impression that Teomim's sermon was more than a politically correct gesture extended to the government and reading public of Frankfurt. On the contrary, the rabbi offered a deeply felt expression of loss for the young prince, whom he called "der Menschenfreund," and "der wahre Herzog," (the true duke), and referred to him as holy, a jewel, an ornament of mankind, and a superhuman ruler.[45] He artfully wove into his sermon biblical passages that succinctly captured the circumstances of the loss, such as Psalm 32:6, "When mighty waters rise, they will not reach him"; Job 30:26, "When I looked for God, evil came"; and Isaiah 57:1, "The righteous perish and no one takes it to heart."[46] He consoled his audience on the loss of this blessed leader by describing his heavenly resting place and the need to preserve his memory by being faithful to the moral obligations he embodied.[47]

Only a few months later, in the summer of 1785, the well-known *Haskalah* figure Naphtali Herz Wessely (1725–1805), writing in Hebrew, devoted a long poem to the death of the fallen prince in the pages of *Ha-Me'asef.* In a note contextualizing the soulful composition, the editors explained the tragic death of Leopold and his selflessness in trying to save his citizens from the ravages of the flood. Wessely, like Teomim, was unrestrained in expressing his grief in the exalted language of his extended poem. He called Leopold a courageous and pious man who would live on in the heavens, a beloved friend, and one who had done incomparable good to others. He underscored the sense of loss the entire community of Frankfurt and beyond now felt, and the great injustice of losing so young a man in his prime. The goodness Leopold accomplished, Wessely noted, was not limited to any particular community but was freely given to Jew and Christian alike. Leopold was a "stone of help" (*even ha-ezer,* a title commonly associated with Jacob ben Asher's legal code) to all of humanity, and all creatures mourned his passing. Accordingly, "like an angel he ascends to the heavens; you have deserted your earthly home and risen to your divine one."[48]

Considered together with Hurwitz's account of Leopold, Teomim's eulogy and Wessely's poem provide a sense of the impact the prince's death had on the contemporary Jewish community. These two sources appear to be independent compositions unrelated to any particular ideology or cultural agenda, in contrast to Hurwitz's more focused use of the story. Teomim, like Hurwitz, evoked Leopold's standard appellation as a *Menschenfreund*, suggesting that he too may have been aware of the spiritual aura associated with his humane activities. Wessely, too, emphasized that the prince's altruism transcended any religious boundaries separating Jews from Christians.[49]

IN SEARCH OF OTHER SOURCES OF HURWITZ'S UNIVERSAL MESSAGE

Hurwitz offers few other clues about the background of his discourse on neighborly love as meaningful as the story of Leopold. The chapter does contain several citations of ancient philosophers–among them Socrates, Diogenes, Chilon, and Bias[50]—who offer pithy articulations of the moral points Hurwitz was making. One possible source for these quotes, as Resianne Fontaine shows, is Cicero's *De Officiis*, although it was certainly not his major source of inspiration for this chapter.[51] His reference to Bias's pronouncement on loving one's enemies seems to be taken from Diogenes Laertius's *Lives of Eminent Philosophers*,[52] though he may have taken this quotation, as well as the others, from a more recent work still not identified.

We might mention in this regard Hurwitz's rendering of the well-known story of Damon and Pythias in the fourth discourse of the second book of *Sefer ha-Brit*. While not included in the chapter on neighborly love, it is linked thematically in its dramatic portrayal of the meaning of true friendship. One is tempted to consider a relationship between Hurwitz's version of the story with that of Friedrich Schiller, who composed a ballad based on the medieval version of the story in 1798 called *Die Bürgschaft* (The Hostage). In both Hurwitz's and Schiller's versions, Damon rather than Pythias is the character sentenced to death. In the end, however, such a relationship seems unlikely, since Schiller's poem did not appear until 1799, while Hurwitz already included the story in the first edition of his book published in 1797.[53] In Schiller's original version, too, the main character was called Moerus; only in 1804 did Schiller rework the story using the name Damon. Nevertheless, the fact that both Hurwitz and Schiller were inspired by the same story, especially in light of their common interest in the heroics of Leopold, should not be overlooked. It may provide another small indication of a cultural background common to both authors.

There remains one final passage in "Ahavat Re'im" that holds a potential clue about the chapter's origin. This is Hurwitz's unusual commentary on the two cherubim connected to the Ark of the Covenant as described in Exodus 25:17–21: "You shall make a cover [*kapporet*] of pure gold. . . . Make two cherubim of gold—make them of hammered work—at the two ends of the cover. Make one cherub at one end and the other cherub at the other end; of one piece with the cover shall you make the cherubim at its two ends. The cherubim shall have their wings spread out above, shielding the cover with their wings. They shall confront each other, the faces of the cherubim being turned toward the cover." Hurwitz's rendering is found near the end of the chapter and functions as a final artistic homily on the general theme of the chapter.[54]

Hurwitz understood the two cherubim to be children with infantlike faces. In this he followed the interpretation of the Talmudic text as well as that of Rashi. The rabbis interpreted the word *k'ruv* (cherub) as deriving from the Aramaic *ravia*, "a child." K'ruv would then be equivalent to *k'ravia*, "like a child," or "childlike."[55] And Rashi explained: "Cherubim—he shall make figures of [= human beings] in it."[56] Isaac Abravanel, in his biblical commentary of the early sixteenth century, even added that the two cherubim facing each other were expressions of "brotherly love of one for the other."[57] These traditional interpretations also seemed to influence the iconography of Torah valences of late-eighteenth-century synagogue arks.[58]

But Hurwitz departed from his Jewish sources; as he wrote:

> This [Exodus 25:17–20] also comes to teach regarding the strengthening of neighborly love to which it alludes. For it is permanent and embedded in the heart of all human beings in nature from their childhood. For observe, my brother, the way in which a young infant who is given a nice toy entertains himself with whatever it is. But suddenly if one places another infant like him next to him, he will toss away the toy and approach the child, grasping him with his hand and evoking a greater joy in him than any other delight that he had previously possessed. Similarly, the second child will play with the first, enjoying the other while raising their varied voices toward each other with a cry of joy and great gladness. And if the second child is not available to him, the first will choose to make an image of the face of a child out of a piece of small wood around which he will wrap a wool or cotton cloth and place a hat on his head to make it look like a small baby. He will then embrace it and it will make him happier than all the other things that are precious to him. So this is truly a natural sign that the love of neighbors is made permanent and implanted in the hearts of all human beings from their earliest age. Thus the image of the faces of the cherubim hints and reminds us of this sign.[59]

Hurwitz acknowledged that this interpretation of the cherubim was only one of many explanations of these powerful symbols, yet for him, this was the most accessible and the most significant one. Thus the two cherubim indicate humans' need for one another, since "it is not good for man to be alone" (Genesis 2:18) and because all human beings in society have a natural inclination to seek one another out. While the Ark cover represents the commandments involving relations between man and God, the cherubim point exclusively to relations among human beings, and these take priority over all divine commandments.[60]

While Hurwitz's interpretation rests firmly on traditional Jewish sources, his account offers a fresh and original slant, with its two infants seeking each other's company in so intuitive a fashion. The childlike angels may be derived from rabbinic sources, but they also suggest those *putti*-like creatures featured in Christian art of the Renaissance and Baroque. One also wonders whether the images of affectionate childlike angels linked to each other, embracing their common humanity, might be located in the era in which Hurwitz worked on this chapter. The Freemasons were especially fascinated with everything having to do with the Temple and the Ark of the Covenant, including the cherubim themselves. If indeed our hypothesis about Hurwitz's awareness of Freemasonry has any credibility, might one imagine that he visualized the cherubim in sources—oral or iconographic—found among Masonic circles of his day?[61]

In the end, our attempt to contextualize this and other passages in Hurwitz's discourse on neighborly love must remain speculative. The story of Leopold remains the most intriguing source in relating Hurwitz's ruminations to a larger community of scholars committed to cosmopolitan ideals. Yet regardless of the uncertain identification of those sources, what remains unambiguous is his effort to promote such an original and bold agenda in the first place. And even more noticeable is the link that Hurwitz establishes between loving all of humanity and imbibing the Divine Spirit, that is, in becoming a prophet. Only by embracing all human beings is divine revelation possible. Accordingly, Hurwitz's moral cosmopolitanism is not simply a social theory or even a philosophical commitment. It is much more. It declares that a love for humanity on the part of Jews is at the very core of their religious consciousness. So striking a formulation suggests no less than a radical redefinition of the Jewish faith, a harsh social criticism of the author's immediate community, and a call for a new ethic transcending national and ethnic boundaries. Only a person capable of loving and appreciating others can reach the most elevated level of spirituality and prophetic insight and so attain Judaism's highest ideals. Whether or not Hurwitz or his readers fully grasped the import of his message, it appears in hindsight that he had indeed articulated a radical understanding of Judaism for his own time and perhaps for ours as well.[62]

CHAPTER 6

The Readers of *Sefer ha-Brit*

The most remarkable feature of *Sefer ha-Brit* was its popularity. Although other Hebrew compendia of scientific information were published during Hurwitz's lifetime, none were even remotely as successful. Hurwitz's work was published in some forty editions between 1797 and 1990, including three abridged Yiddish translations, seven in Ladino (including three of only one section), and fifteen Hebrew versions in Warsaw alone.[1] In contrast, the scientific works of two of Hurwitz's contemporaries, from which he clearly copied, had but limited circulation. The first part of Barukh Lindau's *Reshit Limmudim* (The Beginning of Studies) was published only three times, and the second part only once. A complete edition appeared only in Lemberg in 1869, twenty years after the author's death.[2] Barukh Schick's *Ammudei ha-Shamayim ve-Tifferet Adam* (Pillars of Heaven and the Splendor of Man) was published only once, in Berlin in 1776–77.

A careful review of all the editions, translations, and abridgments of *Sefer ha-Brit* (see appendix 1) offers little insight into the reasons for the book's popularity. Some have assumed that the book sold so well because the author cleverly presented it as a commentary to Ḥayyim Vital's *Sha'arei Kedushah* so that it could be appreciated simultaneously as a compendium of science as well as a work of kabbalistic *musar* (ethics) or even a spiritual journey to imbibe the Holy Spirit. Information about new scientific discoveries was thus presented within a pious discourse acceptable to traditionalists, who would also find Hurwitz's attack on philosophy, Sabbateans, and certain Maskilim praiseworthy. Others have mentioned the initial confusion over the identity of the author, who hid his name in the first edition. Hurwitz himself later mentioned that some had attributed the work to Moses Mendelssohn, while others claimed that Elijah the Gaon of Vilna was the author. Such a confusion might well have encouraged book sales. Hurwitz's aggressive salesmanship may also have contributed to the success of the book.

Such explanations, though plausible, hardly offer the full story of why and how *Sefer ha-Brit* became one of the most popular Hebrew books published in the modern era; why it was the primary means by which traditional Jews learned about the natural world; and why it was read and cited by Jews of diverse regions, cultural backgrounds, and ideologies. For the historian attempting to understand its reception across so wide a spectrum of readers, it is necessary to examine carefully when and how it was cited, who were its admirers and critics, and how it was elevated as a standard authority by all kinds of Jews well into the twentieth century. I have collected in this chapter many of the references to *Sefer ha-Brit* that I have identified to this point. Although I expect there are many more yet to be discovered, this initial probe already offers a rich sampling of testimonies on how the book was read and comprehended and provides at least tentative answers as to why the book was so popular. More significantly, it provides a deeper understanding of how Jews of many persuasions of the last two centuries, but especially those in the traditionalist camps, absorbed, wrestled with, and ultimately learned to tolerate and even approve of the findings of the new sciences. The reception history of this composite book of science, kabbalah, and ethical guidance offers no less than a major case study in the process of modernization of large segments of world Jewry. What follows is an accounting of some of the most interesting examples of Hurwitz's readers.

THE FIRST READERS

The first significant reaction to the publication of *Sefer ha-Brit* appeared in the primary literary forum of the German *Haskalah*, the journal *Ha-Me'asef*, some eleven years after the book was first published in Brünn in 1797.[3] It was in fact the only occasion when the journal devoted so much space to reviewing a single book. The author was probably Aaron Wolfson-Halle (1754–1835), originally the editor of *Ha-Me'asef*, who had previously displayed strong interest in the natural sciences.[4] Whoever he was, the reviewer initially praised the book, which he knew only from the first edition, not the two others that followed, including the pirated one: "For he [the author of *Sefer ha-Brit*] has brought honor to his people and precious splendor to our dear language, enlarging its utterances with the words of science and intelligence; and above all else his moral rebuke which he wrote regarding the love of neighbors stands out for which all readers will find contentment and joy."[5] Already worthy of note was his focus on the significance of Hurwitz's work in enhancing the revival of the Hebrew language, along with his appreciation of the book's progressive attitudes toward the non-Jew, two issues of primary importance to the readers of *Ha-Me'asef.*

Yet following this general praise, the reviewer offered specific criticisms of the work, focusing primarily on inelegancies of Hebrew expression and grammatical mistakes, for which he proposed corrections, especially regarding medical and scientific terms. While the reviewer did not know the author personally, he said, he knew other scholars who were acquainted with him. He related that the author came from Vilna, though he was clearly not the famous rabbi of the city, Elijah the Gaon. Instead, "this man has traveled for the last ten years by way of our city and other cities in Germany and in other countries selling the fruit of his vision [his book]. He is attractive, with a good appearance; all his words sound pleasant, while his countenance testifies to his wisdom."[6]

While this comment implies nothing disparaging about Hurwitz per se, a marginal note that is hardly complimentary was appended about the practices of contemporary Jewish authors:

> For this our heart grieves, over the abasement of the sages and writers of our faith because of our many sins. For the wealthy donors who also love and buy new books do not request them from booksellers but seek them directly from the authors themselves so that the latter go begging from door to door like peddlers bringing the first fruits of their thoughts to everyone's house. Subsequently, the number of book dealers has decreased and the dishonor of the Torah and learning has increased. It is also an embarrassment for us in the eyes of other nations who very much honor their own scholars.[7]

Since the note was tied to the reviewer's comment about Hurwitz wandering through Europe for ten years, he seemed to be criticizing Hurwitz indirectly, or at least the donors to whom he turned in order to sell his books. This is at once an interesting comment on the passion of rich Jews to purchase their books directly from the authors, to the detriment of the conventional book dealer, and an explanation of Hurwitz's constant wanderings, his cultivation of rich patrons, and the ultimate reason why his book was so successful. He had invested enormous time and energy in peddling his book. Given his own remarks about selling the book and his German letter to the Prague censor Karl Fischer (see chapter 2), Hurwitz appears to fit precisely the portrait of the author who hawks his books to the wealthy and dishonors himself.

Still other criticisms followed. The reviewer pointed out that the book was not entirely original, mentioning at least one of its sources, Barukh Lindau's *Reshit Limmudim*.[8] Although he acknowledged that the book often did a good job of presenting new scientific material, calling it an encyclopedia, as the author did, was a misnomer, since the book was not organized in a logical manner, alphabetical or otherwise.[9] Rather, it was entangled in kabbalistic mysteries for which the reviewer had no sympathy. In an era when mystical fantasies were declining, so he believed, it was a pity the author could not present the substance of the book without such impediments. Hurwitz's inability to accept Copernican cosmology was another weakness of the book.[10] In the final analysis, from the perspective of this reviewer, himself a promoter of the Jewish enlightenment, *Sefer ha-Brit* did not fulfill its promise as a scientific encyclopedia. Ironically, it appears in retrospect that the reviewer's specific points of criticism—especially the manner in which the book was sold and the fusion of science with kabbalah—were precisely what led to the book's great popularity.

Some seven years later, in Breslau, a relatively unknown scholar named Moses ben Eliezer Phoebus Koerner (1766–1836) published a work he called *Ke'Or Nogah* (As Radiant Sunlight [as in Proverbs 4:18], 1816). The rest of the title is also worth noting: *Shaviv ve-Niẓuẓ ha-Maḥberet Or Nogah: Ve-Gam Hakimoti et Brit ha-Rishonim Lo ke-Katuv al Sefer ha-Brit* (A Spark and Glitter of the Composition "A Radiant Sunlight": And Also I Have Erected the Covenant of the First Rabbis Not as It Is Written on *Sefer ha-Brit*). In other words, Koerner offered his work as a kind of preview of a larger work that he hoped to publish later (but never accomplished). And one of the major incentives in composing this modest work was to challenge some of the thoughts articulated in Hurwitz's *Sefer ha-Brit*.

Moses Koerner was rabbi in Rendsburg (Schleswig), Shklov, and Grodno. Very much like Hurwitz, he traveled extensively in his later years to promote his own primarily homiletical compositions. He eventually settled in Breslau, where he died. He was a descendant of the sixteenth-century rabbi Yom Tov Lipmann Heller, whose autobiography, *Megillat Eivah* (Scroll of Enmity), he published in 1837, together with notes and a German translation.[11]

As we saw in chapter 2, Koerner knew Hurwitz personally, referring to him as "an exalted friend, my close friend, the perfect wise man, investigator and fearer of God." They had met at the home of Ẓevi Hirsch Levin (1721–1800), the chief rabbi of Berlin, who was also an admirer and supporter of Moses Mendelssohn, Barukh Schick, and other members of the enlightened circle surrounding Mendelssohn. Hurwitz had gained the rabbi's confidence, and Levin later wrote a letter of introduction to his brother Saul Loewenstamm, the rabbi of the Ashkenazim in Amsterdam, when Hurwitz set out for the Netherlands. Loewenstamm subsequently wrote one of the approbations to *Sefer ha-Brit*. Koerner's testimony further confirms Hurwitz's close connections to the Berlin Maskilim. From their first meeting, Koerner says, "I loved him with an eternal love and I knew him by name and I came with him into the bond of the covenant [Ezekiel 20:37]."[12]

Despite these close bonds, Koerner did not hesitate to criticize the work of his friend in print. He did so without hatred, jealousy, or need for self-aggrandizement, he noted, mentioning the mixed review of *Sefer ha-Brit* in *Ha-Me'asef* as precedent and pointing out that Hurwitz himself had confessed his own limitations in challenging previous scholars.[13]

When one wades through his lengthy rebuttal of Hurwitz, however, one senses a certain disingenuousness in his challenging Hurwitz in so public and dramatic a manner. Moreoever, it seems that Koerner was overstating his case regarding the deficiencies he had supposedly unearthed. In the first place, Koerner never addressed the major parts of *Sefer ha-Brit* but only a single section on the love and fear of God. For someone reading his criticism alone, it would have been impossible to know that the book encompassed an extensive description of the natural world, a social critique of Jewish-Christian relations, and more. Secondly, his repeated claim that Hurwitz misconstrued the relation between love and fear of God and had misinterpreted several rabbinic passages seems rather insignificant in contrast with the major themes of the book.

What irked Koerner the most, apparently, was Hurwitz's critique of his hero Moses Maimonides and his assault on the project of medieval Jewish philosophy in general: "And thus I have sought truth, and truth has sustained me, since it is my rock, my deliverance, and my stronghold. And I will trust it in defending the righteous one, our master the Rambam, not as the covenant [i.e., *Sefer ha-Brit*] did to destroy the foundation [that Maimonides] had established in his *Sefer ha-Madah* with knowledge, understanding, and wisdom."[14] And elsewhere, he writes: "May Heaven be my witness that all of my labor which I have exerted was to demonstrate the splendor of our teacher Moses [Maimonides] to overturn the perverse comments and the slanderous talk of *[Sefer] ha-Brit*."[15]

Koerner had indeed correctly noted that Hurwitz's book represented a challenge to the Maimonidean tradition and to the place of philosophy in Judaism. He never mentioned, however, Hurwitz's embrace and elevation of the kabbalah in place of philosophy. Nor did he acknowledge the larger context and scope of his antiphilosophical arguments, which extended far beyond this single chapter. Koerner repeatedly emphasized to his readers that his argument against Hurwitz was neither personal nor about jealousy but only about the love of truth. One wonders, however, whether he wasn't pleading too hard and whether personal considerations might have fueled his frontal, albeit unsuccessful, attack. He had correctly assessed the critical comments Hurwitz had made regarding Maimonides, but he failed to appreciate the larger significance of this critique in the context of the book's overriding message. In focusing on only one small part and offering arguments that were superficial and unconvincing, he genuinely misrepresented the content of the book as a whole. Perhaps this explains the negligible impression *Ke'Or Nogah* made and Koerner's failure to publish the larger composition that he so dramatically announced in this modest publication.

SEFER HA-BRIT IN THE WRITINGS OF MASKILIM

Despite the two early mixed notices of *Sefer ha-Brit*, the book soon drew favorable comment from other writers, most notably from Maskilim or those sympathetic to the ideals of the Jewish enlightenment (*Haskalah*). In 1820, the Hebrew writer and educator David Caro (c. 1782–1839) of Posen published a strong rebuttal to the pamphlet *Eleh Divre ha-Brit* (These Are the Words of the Covenant, 1818), compiled by some of the leading Orthodox rabbis of Europe to protest the innovations of the Reform Temple in Hamburg. Three years later in Vienna, one of the leading figures of the Galician *Haskalah*, Judah Leib Miesis (1798–1831), republished the second part of Caro's book under a new title, *Sefer Tekhunat ha-Rabbanim* (The Book on the Character of the Rabbis). It was a scathing indictment of the traditional rabbinic leadership of the day. In a section insisting that the rabbi should teach Jews to love all of humanity, Miesis added the following note:

> The author of *Sefer ha-Brit* is exacting in his honorable discourse entitled "Ahavat Re'em" in saying, "If it was referring explicitly to Jews in this verse

> [Leviticus 19:18], it would say, 'And you shall love your brother,' as it is written in reference to usury: 'You shall not take interest from your brother [Deut. 23:20].' Or it might say, 'And you should love your countrymen' as is it written: 'You shall not take vengeance or bear a grudge against your countrymen [Leviticus 19:18],' but the intention of 'to your neighbor' is to indicate a person like you, one who participates in the building of society [*osek be-yeshuvo shel olam*]."[16]

The reviewer in *Ha-Me'asef* had mentioned in passing this section on universal morality, but Mieses was the first to praise it unequivocally and to cite it to bolster his own argument.

The Caro-Miesis work caused quite a sensation among traditionalists, who furiously defended the rabbis against their radical critics. One of the most outspoken polemicists was the Ḥasidic leader Nathan Sternharz of Nemirov (1780–1845), the chief disciple of the Ḥasidic rabbi Naḥman of Bratslav (1772–1811).[17] We shall return to Sternharz and Naḥman later in this chapter. Suffice it to say here that both men admired *Sefer ha-Brit* as well, and although they were silent regarding Hurwitz's position on universal ethics, they utilized other sections of the book that were more to their liking. In the same vein, Caro's initial critique was surely directed against the orthodox rabbi Moses Ḥatam Sofer (1762–1839) of Pressburg, Hungary, who had led the attack against the reformers. Ironically, Sofer provided one of the strongest endorsements of Hurwitz's work, as we shall see, surely encouraging a wide readership within orthodox circles. Although Sofer and Sternharz stood in complete opposition to the radical Caro and Miesis, these figures nevertheless shared a great appreciation of *Sefer ha-Brit*. Such examples already provide great insight into the work's popularity across a broad spectrum of the Jewish community. For each of these factions, despite their disparate ideologies, the book held a special appeal.

Isaac Beer Levinsohn (1788–1860), often considered the "father" of the *Haskalah* in eastern Europe, was also familiar with *Sefer ha-Brit* and cited it several times. In his major manifesto *Te'udah be-Yisra'el* (A Testimony [or Warning] in Israel), he focused exclusively on Hurwitz's critique of the exalted place that philosophy had assumed in Jewish culture. He challenged Hurwitz's interpretation of the Polish rabbi Moses Isserles and then brought a quotation from the medieval philosopher Joseph Albo, also cited by the *Ha-Me'asef* reviewer. The point of this criticism was to defend the integrity and legitimacy of the rational pursuit of knowledge in Judaism against those who would deny or limit this path. Levinsohn stressed not Hurwitz's enchantment with the study of the natural world but his engagement with kabbalistic mysteries.[18]

Elsewhere, Levinsohn offered a page-by-page criticism of another contemporary work, *Ha-Torah ve ha-Pilosofia* of Isaac Samuel Reggio (1784–1855), an Austro-Italian scholar also committed to the ideals of the Jewish enlightenment. Among other things, Levinsohn objected to Reggio's conclusions based on his faulty reading of *Sefer ha-Brit.* As in his earlier work, Levinsohn focused on the section of Hurwitz's work that criticized philosophy. Reggio first made the unsubstantiated claim that Hurwitz relied on the classic work of Isaiah Horowitz (c. 1565–1630) *Shnei Lukhot ha-Brit* (The Two Tablets of the Law) for his inventory of antiphilosophical statements in Judaism. Horowitz, in the estimation of Levinsohn, was not antagonistic at all to the study of the sciences, including philosophy. Levinsohn also pointed out Reggio's embarrassing mistake of confusing the medieval thinker Joseph Yabeẓ with the eighteenth-century Jacob Emden (also called by the acronym Yabeẓ). In any case, neither was antiphilosophical, and Reggio was misled by relying on the text of *Sefer ha-Brit.* For Levinsohn, accordingly, Hurwitz emerges as a a one-dimensional critic of rationalism and nothing more.[19]

Samuel Joseph Fuenn (1818–1890) of Vilna belonged to the next generation of Maskilim and situated himself more firmly within their traditionalist camp. Fuenn is important in the tracing of the legacy of *Sefer ha-Brit* because he was the first and probably only scholar to underscore Hurwitz's Vilna roots. He devoted a long entry in his history of the Jewish notables of the city, *Kiryah Ne'emanah* (Faithful City, 1860), to the "famous rabbi" Pinḥas Hurwitz.[20] Fuenn gleaned most of his information from the two introductions to *Sefer ha-Brit*, of which he consulted the Vilna edition. Although he mentioned the names of Hurwitz's parents, Meir and Yenta, he had nothing else to say about them. Fuenn also referred to the common confusion over the author's identity as either the Gaon of Vilna or Mendelssohn. Such a characterization might have provided the well-balanced mixture of enlightenment and traditionalism with which Fuenn would have been comfortable.

While recognizing the strengths of Hurwitz's book, Fuenn also mentioned its limitations. He baldly stated that Hurwitz had probably exaggerated his accomplishments and that his knowledge of the sciences was often superficial. In this context, he cited extensively from the critical review in *Ha-Me'asef*, referring obliquely to Hurwitz's untenable position regarding Copernicus, although he still placed him in the same exalted company as such Hebrew scientific writers as Joseph Delmedigo and Israel Zamosh. Fuenn wound up his biographical sketch by alluding to Hurwitz's essay on universal ethics and to other works, mostly unpublished, described in *Sefer ha-Brit*. He ends by claiming him for the glory of Vilna and its sages, "to whom our community gave birth and nourished for its honor and praise."[21]

Eliezer Zweifel (1815–88), a contemporary of Fuenn, was best known for his twenty-year association with the Zhitomer rabbinical seminary, for his moderate maskilic views, and for his defense of Ḥasidism, a stance that evoked the scorn of many of his more radical contemporaries in the Jewish community. Zweifel's citation of *Sefer ha-Brit* is found in the last work he wrote, *Sanegor* (Defense Counsel), a response to the missionary Alexander McCaul and his assault on the Talmud, published in Warsaw in 1885. Zweifel counters McCaul's deprecation of rabbinic ethics by providing a comprehensive inventory of biblical, rabbinic, and contemporary Jewish sources on Judaism's ethical teachings regarding women and non-Jews. It is in the latter context that Zweifel found Hurwitz's moral cosmopolitanism important. He first cites him in a series of quotations about the loyalty of Jews to their governments. Then he turns to a catalogue of sources that treated "universal love," stressing the love of humanity in Judaism. These include such modern writers as Moses Gentili, Eliezer Fleckeles, Barukh Jetteles, Israel Landau, Jacob Emden, and many others not commonly associated with such a universalist position. Here Hurwitz's chapter "Ahavat Re'em" is cited, but not directly. Instead, Zweifel strangely quoted the Bulgarian rabbi Abraham Alkalai (1750–1811), who in his halakhic work *Zakhor le-Avraham* (Remember Abraham) had himself quoted Hurwitz, "the divine scholar and famous sage," as early as 1798, soon after the initial publication of *Sefer ha-Brit*.[22] Zweifel, like Alkalai, Miesis, and the reviewer in *Ha-Me'asef* before him, singled out the ethical position propounded by Hurwitz and praised it.[23]

ḤATAM SOFER AND *SEFER HA-BRIT*: ACQUIRING AND MAINTAINING THE SEAL OF RABBINIC AUTHORITY

Given Moses Ḥatam Sofer's special place as the primary leader of the so-called ultra-Orthodox camp in the first half of the nineteenth century, and especially his prominence as an enemy of reform, it comes as somewhat of a surprise to discover his embrace of the study of the sciences in general and of *Sefer ha-Brit* in particular.[24] But here is the apparently reliable testimony of his grandson, Solomon Sofer (1853–1930), from his *Ḥut ha-Meshullash* (The Triple Cord, originally published in 1887), a collection of biographies of famous members of his family:

> He [Moses Sofer] was an expert in the science of measurements, algebra, and the calendar. In this he was quite efficient and able. . . . He was also proficient in astronomy and physics and began to compose a special book on these sciences so that it would be handy to his students to study from it without having to consult gentile authorities. But when *Sefer ha-Brit* was published, he examined it from beginning to end and declared that anyone who desires to satisfy himself in various sciences should acquire it: "For I and you are indebted to the author, since I was able to acquire much valuable time in not having to write a book like this for you."[25]

Sofer never completed the "special book," although an apparent fragment of it was published in one of his responsa.[26] One wonders, in any case, whether his endorsement of Hurwitz's book encouraged other traditionalists to buy it. It certainly would have pleased Hurwitz to receive it prior to publication.

Hurwitz, as we have seen, sought out seven other rabbis to officially approve of his book when it was published in 1797, but Sofer was not one of them. Although Hurwitz had lived in Pressburg for some time in the home of Beer Oppenheim, probably in 1794, Sofer, who became the chief rabbi only in 1806, was not yet living in that city.[27]

There remains one more piece of evidence, recently discovered, to suggest a relationship between Sofer and Hurwitz in subsequent years. In the first edition of *Sefer ha-Brit*, Hurwitz wrote the following at the end of his description of the sun and moon:

> Behold, many people will never know or understand the course of the sun and moon and their circuit as we have described it. They rather imagine that when the sun shines each morning, it appears to arrive from above the firmament, from its other side, and then it exits through the gate of heaven, through the window eastward, piercing the transparent and opaque windows in the firmament. . . . They think that the lights [the sun and the moon] enter through windows to illuminate the earth and its inhabitants. But if you explain to them their true course, they will not comprehend, given the weak nature of their minds. They thus appreciated the formulation of the ancient sages in the prayer "All shall thank you," expressed

in simple human language and as they had imagined it. . . . Thus they declared: "[God] who opens every day the doors of the gates of the East, and cleaves the windows of the firmament, leading forth the sun from its place and the moon from its dwelling, giving light to the whole world. . . . " Thus they were obligated to compose the prayer according to their [the common people's] understanding even if in fact it isn't true, so that they would sing praises of God's ways in a manner consonant with their hearts. . . . Know accordingly that one who wants to compose a good prayer need not be precise in its formulation to speak the truth, since the masses and most people will not understand anyway. But he alone should be aware that the message is a truthful one.[28]

Sofer noticed this passage and was clearly annoyed. Writing in 1799, he said: "Understand correctly and be angry at the author of *Sefer ha-Brit,* who was mistaken in what he wrote in the fourth section, chapter ten, beginning with the words 'Behold' etc., and may God forgive him."[29] For Sofer, the words of the sages were by definition true; thus it was incumbent upon Hurwitz to understand the prayer's formulation as pointing to a higher truth, a spiritual one, and not as a falsehood written only to seduce the masses. And apparently Hurwitz got the message. In the revised edition of his book, a long addition to chapter ten appears, beginning thus: "While the ancient sages who wrote this wonderful poem, speaking in the language of common people, presented it to fit their understanding, nevertheless they themselves knew that these words were very reliable according to the secret meaning of the hidden place of God in Heaven high above the solar sphere. In that place there are actually windows for the course of the spiritual sun to enter the spiritual windows that are there. . . . If we pay attention to understand the words of the sages and the opinion of the holy ones who speak of this matter, we will see wonders." For Hurwitz, in this new version, both the scientific explanation and the spiritual one are valid and do not contradict each other, since "these and these are the words of the living God."[30]

What follows is a long list of rabbinic sources that speak credibly of the heavenly windows. If there remained any doubt that Hurwitz revised his chapter according to Sofer's wishes, one of these sources surely confirms this impression. Immediately before his criticism, Sofer wrote the following in discussing the location of the center of the earth: "This matter is impossible to determine by rational proof or by the testimony of the Torah; it can only be established based on our tradition from the oral law. But look what the author of the *Asarah Ma'amarot* [the kabbalist Menahem Azariah da Fano, 1548–1620], *Ma'amar ha-Etim*, section 9, wrote on the windows of the firmament and understand it well."[31] Hurwitz cites the same passage but argues that Fano's simple explanation of the windows of heaven based on observable data is not tenable either. He ends by mentioning the discussions about the heavenly windows of two other kabbalists of whom Sofer would have certainly approved, Solomon Aviad Sar Shalom Basilea (c. 1680–1749) and Jacob Emden (1697–1776).

This fascinating exchange offers a rare glimpse of Hurwitz's attentiveness to the opinion of so significant a rabbinic scholar as Ḥatam Sofer. His additions in the new version of his book were not random but were calculated to meet the needs of Sofer and the orthodox readers who would buy his book. In the interest of a pious science, Hurwitz was forced to retreat from his first formulation, although, to his credit, he did not succumb to Sofer's wishes altogether, but rejected the pseudo-scientific conclusions of Azariah da Fano.[32]

Other close associates of Sofer also consulted *Sefer ha-Brit*. Akiba Eger (1761–1837), Sofer's father-in-law, who was also later identified with the orthodox camp of rabbis opposing reform, clearly viewed Hurwitz's work as a source of scientific knowledge regarding the heavens and cited it in his running commentary on the *Shulḥan Arukh*.[33] Akiba Yosef Schlesinger (1837–1922), a disciple of Sofer's and a champion of ultra-orthodoxy in Hungary, published the Hebrew text of Ḥatam Sofer's last will and testament along with the appendix he composed entitled *Na'ar Ivri* (Hebrew Youth). In this discourse, a passionate attack on reform and acculturation, he cited Hurwitz's criticism of the medieval philosophers and his assault on the radical Maskilim and those who pursue the philosophical path over that of the kabbalah.[34] While Eger saw Hurwitz as a source of knowledge about the natural world, Schlesinger saw him as a comrade in arms against enemies of the Jewish faith.

WORMS UNDER THE MICROSCOPE: *SEFER HA-BRIT* AND THE HALAKHIC DECISORS

Moses Sofer's endorsement of Hurwitz's compendium was indicative of a general trend that remained constant for almost two centuries: a reliance on the book as an authoritative source for rabbis to consult in determining

Jewish law. A wonderful example is Hurwitz's discussion of vinegar in *Sefer ha-Brit*:

> All vinegar from wine and beer or other liquids is filled with very small and fine worms that cannot be seen by the eyes alone. However, if one places the vinegar in a clear flask by the window so the sun shines on it, one can recognize them naturally with a good eye without the need of a magnifying glass. . . . The reason for this is that vinegar cannot be created without the drink being spoiled first and becoming moldy until it ferments. And all moldy things have worms in them. Thus, it will not help if one strains it, since through every cloth or any other strainer the worms will pass together with the vinegar, because almost the entire essence of vinegar is made up of fine worms that live in it. . . . Thus one cannot drink vinegar until it has been boiled to kill the worms, and then it is useful to strain.[35]

Hurwitz's reference to what is now called *Turbatrix aceti* (vinegar eels or vinegar nematodes) was first noticed by Abraham ben Yeḥiel Danzig (1748–1820), a merchant and later *dayan* (rabbinical judge) of Vilna. In *Ḥokhmat Adam: Binat Adam* (The Wisdom of Man: The Intelligence of Man), his halakhic discussion of the *Yoreh De'ah* and other sections of Joseph Karo's Jewish law code, the *Shulḥan Arukh*, he was asked to comment on "what was written by one scholar in his book *Sefer ha-Brit,* who maintained that vinegar cannot exist without worms arising in it. One observes the vinegar filled with worms through a glass instrument called a microscope."[36] Accordingly, it is prohibited to drink the vinegar unless it is first boiled and then strained, since straining alone is insufficient.

Danzig seemingly was not familiar with Hurwitz, but he felt compelled to respond to his questioner, who had somewhat garbled Hurwitz's precise wording. Hurwitz called for screening the vinegar by the naked eye through a clear flask placed in the sunlight. Although he mentioned the microscope later in the chapter, Hurwitz did not call for its use in this instance, an important distinction that was lost on Danzig. Danzig disagreed with what he thought Hurwitz had said, arguing that what is seen by the naked eye is sufficient in deciding what is forbidden and what is permitted. For him, the analogy of the egg and the chicken is relevant: what exists only potentially is not yet the actual thing. Thus, what is under a microscope cannot be considered something that exists in reality. Although the Torah allows the eating of worms, the rabbis prohibit it for the sake of appearances (*marit eyin*). Yet for

Danzig, there is no danger in eating meat or vegetables cooked with vinegar, even for the most pious Jews.

Ẓevi Hirsch Shapira (1850–1913), the Ḥasidic grand rabbi of Munkacs and the author of an important commentary on the *Shulhan Arukh,* referred to Hurwitz's discussion of vinegar several times, clarifying that Hurwitz called for observing the worms with the naked eye and not necessarily through a microscope and accepting his testimony that worms always accompany the fermentation process.[37] Israel Abraham Alter Landau (1886–1942), the rabbi of Edeleny, Hungary, also discussed Hurwitz's stringent position regarding the appearance of worms in vinegar. He first mentioned Shapira's discussion, acknowledging that in the wine vinegar Hurwitz inspected worms were present but stating that that was not necessarily the case for all kinds of vinegar. Having noticed that earlier rabbis did not inspect their vinegar as carefully as Hurwitz prescribed, and based on his experience with the local market, he claimed that the precautions Hurwitz had advised were not necessary and Jews could use vinegar freely.[38] Thus, although neither rabbi adopted the opinion of *Sefer ha-Brit* as the legal standard, it was nevertheless taken quite seriously, with Hurwitz considered an expert to be cited and discussed.

Having entered the legal discussions of worms and fermentation though these previous rabbis' writings, *Sefer ha-Brit* continued to be a reference on the subject throughout the twentieth century. Rabbi J. H. Beck-Cohen of Montreal referred to it in his legal digest *Sefer Beit Shmuel,* published in Jerusalem in 1961.[39] Elijah ben Yosef Avirazal, the sephardic rabbi and head of the yeshiva Ẓuf Devash in Jerusalem, discussed the same issue in 1993, paraphrasing the argument he had read in Rabbi Danzig's responsum but citing Hurwitz's text directly.[40] Former sephardic chief rabbi of Israel Ovadiah Yosef (1920–2013) devoted a lengthy responsum to the subject of viewing worms under the microscope in which he cited both Danzig and Hurwitz. He concluded that the law applies only to what can be viewed with the naked eye, not by the magnifying glass or microscope. Referring again to the difference between an egg and a chicken, the *halakha,* Yosef maintained, cannot take cognizance of subclinical phenomena. Thus, vinegar displaying swarming worms under the microscope is not prohibited from use either by the Torah or the rabbis.[41]

Finally, the Ḥasidic rabbi Menashe Klein (1924–2011) of Brooklyn and Jerusalem, known as the Ungvarer Rav, also referred to the same passage in *Sefer ha-Brit* in one of his responsa. He was asked whether a photo of the Bible whose script is too small to be read except through a magnifying glass or microscope should be considered a holy book. He concluded that a Hebrew Bible is only holy when it can be read with the naked eye. To substantiate his case he cited Danzig's discussion of Hurwitz and other responsa dealing with the same subject of worms under the microscope. He further argued that photographers who diminish the size of the paper and the script often leave off letters and can easily produce a highly flawed work. Expanding the discussion of seeing natural objects to that of seeing the Hebrew Bible under the microscope, Klein arrived at the conclusion that something is holy only when recognized naturally through normal vision.[42]

SEFER HA-BRIT AND THE BRATSLAV ḤASIDIM

While Pinḥas Hurwitz openly attacked the messianic figure of Shabbetai Ẓevi and his followers on several occasions in his book, even offering considerable detail about their antinomian behavior, he was quite silent about Ḥasidic leaders and their followers.[43] One wonders if this silence reflects a certain ambivalence. On the one hand, he shared with them a deep commitment to Jewish spirituality and an appreciation of kabbalistic theosophy and literature. On the other hand, their challenge to the organized Jewish community and its rabbinical leadership no doubt irked him. Or perhaps he was silent because he saw the Ḥasidic community as potential readers of his book and thus considered it prudent to remain neutral in the open struggle between Ḥasidim and their opponents.

Is it possible, however, to detect a veiled reference to early Ḥasidism in this outburst regarding the Sabbateans? "Therefore, people of the God of Israel, be careful with your lives. Don't touch the messiah [Shabbetai Ẓevi] because of his uncleanliness and that of his sect and all of the words and books and formulations alluded to him. Moreover, be careful of any new sect and new associations that have recently emerged among the children of Israel who have separated themselves from the community, turning to other opinions, behaving and following their own path in a way unfamiliar to their ancestors." He was particularly concerned with those who deviated from the path laid out in the *Shulhan Arukh*. He singled out Torah scholars who associated themselves with Shabbetai Ẓevi but added: "Also among the sect of the Nazirites who emerged during the Second Temple period there were men of good deeds and very pious men [*Ḥasidim gedolim*], but nevertheless they left the Jewish fold because they failed to pay attention to the sages of the Mishnah who lived then. They acted with piety [*be-ḥasidut*] according to their own understanding and prompted by their own hearts. . . . Thus be very careful of what is new."[44]

Do the words *Ḥasidim* and *ḥasidut* refer only to pious Jews of antiquity, or did Hurwitz have in mind modern-day Ḥasidim as well? Moshe Idel points to another passage in *Sefer ha-Brit* that he believes contains an explicit attack on Ḥasidic views: "There are [persons]," Hurwitz writes, "who are not interested in secrets and say that everything written in the account of the Creation and in the account of the Chariot, in the *Zohar* and in the writings of Isaac Luria and in those of the ancients, are all parables and metaphors for the powers found in man." For Idel, this passage illustrates the Ḥasidic reinterpretation of Zoharic and Lurianic kabbalistic systems from theosophy to psychology, what he calls "the psychologization of Kabbalah."[45] If he is correct, Hurwitz's oblique references to Ḥasidim might also refer to contemporary ones as well.

Whether or not Hurwitz appreciated the Ḥasidim, some of their leading figures certainly appreciated his book. Some forty years ago, Mendel Piekartz noted that Naḥman of Bratslav and his chief disciple, Nathan Sternharz of Nemirov, had read it carefully and cited it often, referring generally to the "philosophers," "the sages of science," or "naturalists." According to Piekartz, Naḥman found two aspects of the book especially attractive: the appreciation of the wonders of nature, on the one hand, and the critique of the philosophers and their false claims in understanding the meaning of human existence, on the other. In the first instance, Naḥman commended Hurwitz's discussion of lightning rods, which Hurwitz called *Wetterleiter,* and seemed especially taken with the fact that the inventor was killed by lightning as he erected his new invention. Naḥman also appreciated Hurwitz's account of those radical philosophers who depict Moses as a natural magician and his denunciation of them. He cited approvingly Hurwitz's identification of their position with that of the Karaites, a view mentioned by Nathan as well.[46]

Since the time of Piekartz's book, others have expanded the list of references to *Sefer ha-Brit* in both Naḥman's and Nathan's writings. Zvi Mark has noted the theoretical and practical discussions of the role of the imagination in *Sefer ha-Brit* which appear to be absorbed into Naḥman's writings. He also considers Hurwitz's discussions of sleep and dreams as possible sources of Naḥman's thinking. Naḥman also quotes Hurwitz's derisive description of the early Sabbateans (quoted above), though Naḥman was referring not to the early Sabbateans but to the Sabbateans of his own day, namely the Frankists.[47]

To these references one could probably add many more, especially from the writings of Nathan, which have not yet been systematically studied. Two additional references will suffice for now to suggest how influential *Sefer ha-Brit* was on the wider circle of Naḥman of Bratslav and even on one of his chief adversaries. Joseph Perl (1773–1839), the former Ḥasid turned Maskil and a satirist of Ḥasidim, included in his *Megalleh Temirim* (Revealer of Secrets) an alleged conversation a Ḥasid heard in a wine shop. A man not known to the Ḥasid described "the instrument which stops thunder or lightning, I do not remember which. The people there asked him: 'From where did such an invention come?' He responded with his speculations and nonsense so that it seemed to them that these were great thoughts and inventions, while I pretended not to hear anything. When he left, I said to the people who remained that there is no need to search for remote explanations and discoveries, for according to my view, there is a simple explanation. It is know that thunder beats in a place where there is the *kelipah* [the outer "shell" concealing the divine light within all creation—hence, the evil realm in Kabbalah]."[48]

The story goes on to ridicule such populist notions of the Ḥasidim. The relevant part of this story is its discussion of lightning rods, clearly a reference to Hurwitz's treatment of the subject, which was cited by both Naḥman and the radical Maskil Isaac Satanov, whom Hurwitz paradoxically excoriated within the pages of his book.[49] Such a wide circulation of one description of a new scientific invention underscores my earlier discussion of what most appealed to Hurwitz about the new science, namely the thrill of new discoveries, and what appealed as well to his many readers.

The second source is a letter composed in 1807 by the Maskil and merchant Heikel ha-Levi Hurwitz (c. 1750–1822) of Uman to two Karaites named Isaac and Abraham. The writer (later to publish the first modern Yiddish book, a translation of Joachim Heinrich Campe's *Entdeckung von Amerika*), promised to send his two associates the *Phaedon* of Moses Mendelssohn; *Sefer Dod Mordekhai*, the responses of the Karaite author Mordekhai ben Nisan to the Leiden professor Jacob Trigland; and *Sefer ha-Brit,* although the latter "is presently lent to one of my associates."[50] Hurwitz and his merchant son Hirsh Beer had an ongoing relationship with Naḥman of Bratslav. This remarkable source suggests a further circulation of Hurwitz's work traversing not only Naḥman himself but also the circles of an enlightened Yiddish writer and his son and two Karaite intellectuals. Clearly, the kind of pious science that Hurwitz promoted was neither a source of contention nor a subject of suspicion for these readers, despite their disparate ideological positions.

THE EMBRACE OF *SEFER HA-BRIT* BY SEPHARDI INTELLECTUALS

Despite the ashkenazic roots of the author, *Sefer ha-Brit* quickly became a best-seller among Sephardim, both Ladino speakers of the Ottoman Empire and Arabic-speaking Jews from the Middle East and North Africa. Hurwitz, in the opening of the second edition of his book, mentioned his good fortune in selling two thousand copies of the first edition in Poland, Hungary, Germany, Holland, England, France, and Italy but also in "Ereẓ Uẓ, Damascas and Aram, Algiers and Barbaria, and Jerusalem."[51]

As early as 1842–43, Hurwitz's chapter on loving one's neighbors was translated into Ladino and published in Salonika by Yiẓḥak Bekhor Amarachi and Yosef ben Meir Sasson under the title *Sefer Darkei Adam* (The Book of the Ways of Man). They also published another Ladino compilation called *Musar Haskel* (Moral Lesson) that borrows significantly from Hurwitz's work. *Musar Haskel* was republished in Salonika in 1848–49 and 1889–90. In 1846–47, Shabbetai Alaluf and Yiẓḥak Jahon published an enlarged edition of *Sefer ha-Brit,* translated by Ḥayyim Avraham Benvenisti Gateño; it was republished in 1881 in two versions. In addition, the work appeared independently in Constantinople in 1889–1900 in an edition published by Eliyahu Levi ben Naḥmias.[52]

Given the accessibility of the work throughout the nineteenth century in both Hebrew and Ladino versions, it was inevitable that sephardic rabbis would notice it and quote it both as a source of scientific information and for moral instruction. Eliezer Papo (1785–1834), the rabbi of Selestria, Bulgaria, and author of *Pele Yo'eẓ* (Advising Miracle), an extremely popular moral treatise published in Hebrew, Ladino, and even in Yiddish, was the first of a long line of Sephardim to consult *Sefer ha-Brit.* In the first edition of *Pele Yo'eẓ* published in Constantinople in 1824, he refers to Hurwitz's chapter on fear and love as a discussion written "with good taste and intelligence" which serves as a basis for Papo's own discussion. In the Ladino version of the work, published by his son Judah in Vienna in 1870, the reliance on *Sefer ha-Brit* is even more obvious. Papo based his discussions on astronomy, especially his outdated notion of a geocentric universe, on Hurwitz's work. His assault on philosophy and his reference to the philosopher from Königsberg (Kant) were also surely indebted to Hurwitz.[53]

Raphael ben Elijah Kazin (1818–71), the controversial rabbi of Baghdad and later Aleppo and author of three works of anti-Christian polemics, studied *Sefer ha-Brit* carefully and cited it often in *Derekh ha-Ḥayyim* (The Way of Life), published in Constantinople in 1848. Kazin was especially incensed by Christian missionaries who preyed on Muslims and Jews alike. *Derekh ha-Ḥayyim* is a shortened version of a larger work written to defend Judaism against Islam and Christianity and contains a rejoinder to the same Alexander McCaul, the evangelical missionary who had been the target of Eliezer Zweifel, as mentioned above. Kazin constructs a debate between Jacob the Jew, Mati the Christian, and Avad al-Navi the Muslim, all addressing a shepherd who upholds a natural religion.

Why Kazin considered *Sefer ha-Brit* a useful resource for interreligious polemics is not clear. Hurwitz was addressing Jews on how to reconcile science and faith, both those who were ignorant of the sciences and those who had lost their commitment to Judaism in an intellectual world increasingly indifferent or antagonistic to divine revelation. The issue of Judaism's validity in relation to the competing revelations of Christianity and Islam was never his concern. He addressed the Christian "Other" only to argue that Jewish love extends to non-Jews as well. But Kazin's passions were quite different. He read Hurwitz not for his science but for his discussions of Jewish faith, and these he deemed inadequate and flawed. Kazin acknowledged Hurwitz's good intentions but faulted him for not emphasizing the significance of the oral law, a key issue in debates with Christians and Muslims. He was also dissatisfied with his generic explanations of the Jewish faith. Judaism cannot be explained under one rubric, he pointed out, but needs to be presented differently when speaking to a Christian or to a Muslim. He was also unhappy with Hurwitz's presentation of the eternity of Jewish law, charging that his treatment of Jewish sources, especially Maimonides, was lacking.

Kazin observed that Christianity's articulation of the golden rule (Leviticus 19:18) is utopian and unrealistic in comparison with Hillel's superior and more practical formulation, since it is unnatural to love all human beings. He then cited Hurwitz's chapter on loving one's neighbors but strangely ignored its point. In his superficial reading and even misreading of this chapter and many others, Kazin seems to have unwisely chosen the wrong source to fortify his interreligious debates. Hurwitz wrote his book for an entirely different purpose than that required by the Baghdadi rabbi. Obsessed with his need to identify anti-Christian arguments in Hurwitz's work, he could not help but find it deficient and ineffectual.[54]

In contrast to Kazin, Moses ben Raphael Pardo (d. 1888) understood and valued *Sefer ha-Brit* on its own merits. Pardo, a sephardic rabbi from Jerusalem, undertook a journey to North Africa to raise funds for his community. He eventually reached Alexandria, Egypt, where he was offered and accepted the office of rabbi. In an essay entitled "Response Out of Love," published in 1872 in the maskilic journal *Ha-Levanon,* Pardo referred to Hurwitz as "the exceptional scholar, great astronomer . . . kabbalist, treasure filled with all sciences perfect in all fields of learning and ideas."[55] He was especially taken with Hurwitz's discussion of the alien nature of philosophy in Jewish culture and the idea that it was imported by way of the Greeks and Arabs. Pardo was particularly sensitive to Maimonides's philosophical commitments and his identification of philosophical ideas with the secrets of the Torah. But he acknowledged that this identification remained unproven, in contrast to Hurwitz's bold declaration that the Sinaitic revelation, the belief in divine providence, and creation out of nothing were based on sense perception and were unambiguously true. Despite his affection for his medieval hero, the path of Judaism to be followed in this time required a distancing from philosophy.

Even more invested in the thinking of Hurwitz was the famed Iraqi rabbi Ben Ish Ḥai of Baghdad (1832–1909). Ben Ish Hai, the author of a commentary on the Talmud and numerous responsa, was, in many respects, the perfect reader of *Sefer ha-Brit,* given his intense interest in both natural philosophy and kabbalah and in the connections between the two. He considered Hurwitz a legitimate authority on scientific matters and cited him on the place of the Garden of Eden, on the roundness and measurement of the earth, on measuring the weight of air, and on the phases of the moon. He emphasized, like Hurwitz, the speculative, tentative nature of scientific theories, which can be overturned in the wake of new discoveries. Astronomical calculations, he observed, are only meant to be estimations, like those of an architect constructing a building.[56]

He refers to the same discussion of the liturgical poem for the Sabbath that had irked Moses Sofer and prompted Hurwitz to correct it in his second edition. Ben Ish Ḥai devotes a responsum to the same question of how to reconcile the knowledge of science with the cosmological assumptions of the poet. His answer is identical with that of Hurwitz, whom he quotes extensively: literally, the poem makes no sense; it is only meaningful and true as a divine secret.[57]

Isaac ben Abraham Akrish (d. 1888?), a prominent rabbinic judge of Istanbul, appreciated Hurwitz for different reasons. In 1862, Akrish excommunicated the educator Abraham de Camondo for establishing the first modern Jewish school in the city. He insisted that only Ladino and Turkish be taught, in addition to Hebrew, as a concession to the reality Jews now lived in, but he prohibited European languages and culture, saying they would lead to heresy and nonobservance. In his collection of responsa entitled *Kiryat Arba* (Town of the Four [after Genesis 23]), published in Jerusalem in 1876, he enlisted Hurwitz in his campaign to resist the study of foreign languages and culture, calling him "a universal scholar of all fields of knowledge and particularly of philosophy, whose name is better known in Israel than all the ancients." Akrish, who admits that he read *Sefer ha-Brit* in Ladino translation and not in the original Hebrew, praises Hurwitz for his rejection of philosophy and especially for teaching the sciences in Hebrew, accompanied by "the fear of heaven." Although he states that he is unsure he has quoted Hurwitz correctly, not having examined the Hebrew book, he nevertheless assumes that he has captured the essence of his position.[58]

The parade of sephardic luminaries citing *Sefer ha-Brit* continues unabated almost to the present day. The responsa of the Jerusalem rabbi Elijah ben Yosef Avirazal and of the former sephardic chief rabbi of Israel, Ovadiah Yosef, were mentioned above. Another rabbinical judge of the Jerusalem Sephardim and rabbi of the Beit El synagogue, Shalom Hedaya (1864–1945), cited Hurwitz in his homiletical work *Shalom le-Am* (Peace to the People), published in Aleppo in 1896. For him, Hurwitz's pronouncement that parents should educate their children in a specific profession or craft "created a loud noise" and was praiseworthy, although he recognized that not all Jews could fulfill such a high ideal.[59] Another kabbalist of the Beit El circle, Raḥamim David Sarim (1869–1934), quoted Hurwitz in his collection of responsa called *Sha'arei Raḥamim* (The Gates of Mercy), published in Jerusalem in 1926. He lauded Hurwitz's sense of responsibility in publishing works on the kabbalah, particularly since he had novel insights to offer his readership. He was also pleased by Hurwitz's reiteration of the commandment to study the Torah in each of four ways (the simple, allegorical, homiletical, and mystical interpretations of the Bible), called by the acronym *Pardes*, and of the reward that such study brings.[60]

Ḥayyim Blia'h (1832–1919), the rabbi of the Jewish community of Tiemçen, Algeria, in his commentary on the fifteenth-century work of Ephraim al-Nakava (d. 1442) cited Hurwitz extensively regarding miracles. He appreciated Hurwitz's claim that miracles occur, extend to individuals, and fortify the worship of God.[61] Maẓli'aḥ Mazouz (1911–71), a rabbinical judge of Tunis who was murdered by Muslim fanatics, referred to Hurwitz several times in his legal writings on astronomical matters.[62] Finally, Pinḥas Zabiḥi, the rabbinic leader of congregation Ateret Paz in Jerusalem and a one-time associate of Rabbi Obadiah Yosef, quoted Hurwitz at length on the uncertainty of scientific investigation in his *Sefer Tehillat Pinḥas* (The Book of Pinḥas's Praise), published in 2010 in Jerusalem.[63]

MODERN WRITERS AND *SEFER HA-BRIT*

Beyond the rabbinic and maskilic circles drawn to Hurwitz's book, a vast range of Hebrew and Yiddish writers, public intellectuals, and others not so easily classifiable were exposed to *Sefer ha-Brit* and derived great pleasure from its accounts of exotic inventions and discoveries, its moral instruction, and its spiritual journey. Rachel Morpurgo (1790–1871), an Italian Jewish poet, was fascinated with *Sefer ha-Brit*, especially its moral-spiritual message, and referred to it on several occasions.[64] Samuel David Luzzatto (1800–65), the Italian Jewish scholar, poet, and teacher at the rabbinical college of Padua whom we met in chapter 5, cited Hurwitz primarily as a kabbalist and interpreter of the teachings of Isaac Luria and Ḥayyim Vital.[65] Writer Yosef Ḥayyim Brenner (1881–1921) described a character in one of his novels who knew everything about mathematics and the sciences as "not missing anything big or small in the field, from *Sefer ha-Brit* to *Sefer Ayil Meshulash* [a Hebrew work on geometry attributed to Elijah the Gaon of Vilna]."[66] Shmuel Yosef Agnon (1888–1970), the Nobel Prize laureate, recorded Solomon Sofer's comment about his grandfather's appreciation of *Sefer ha-Brit*.[67] Solomon Schechter (1847–1915), a European-born rabbi who served as head of the Jewish Theological Seminary of America, once wrote the following: "It was a dream of my childhood when I learned through the *Sefer haberith* and the letters of Hag Vidaver [Henry Vidaver (1833–82), American rabbi and writer] in the Hebrew weekly *Hamagid* of the existence of a continent on which, according to my simple calculations, people should stand on their heads, and who yet somehow managed to walk erect and free and even move quicker and with a surer pace than we, with all our drill of thousands of years."[68]

The Yiddish writer Isaac Meir Dick (1814–93) read *Sefer ha-Brit;* so did Y. l. Peretz (1852–1915).[69] Aḥad ha-Am (1856–1927), the Hebrew essayist and Zionist thinker, studied Hurwitz's scientific work together with that of the seventeenth-century scholar Joseph Delmedigo.[70] Chaim Tchernowitz (1871–1949), the great historian of Jewish law known as Rav Ẓa'ir, also relates that he read the book in his youth.[71] Isaac Bashevis Singer (1902–91) loved the book and read it together with *Shevilei Olam* (another Hebrew work on geography published by Shimshon Bloch); he adds that Hurwitz's book was special to his mother, who considered it "on a high level."[72]

The Yiddish-language periodical *Forverts* (Forward) has several stories related to Hurwitz's book. In one case, a Frankist keeps *Sefer ha-Brit* in his bag with another book for studying Polish. In another, a fictional rabbi wrestles with his evil inclination, overcoming his doubts by reading about air balloons, beavers, and Kant in *Sefer ha-Brit.* In still another, one reader marvels at the process of male and female pollination described in the book, while a mother sits in bed reading the work for moral instruction.[73]

In a Russian story written by a N. Naumov and translated into Hebrew in the pages of *Ha-Eshkol* in 1902, the author introduces the colorful character of Ḥayyim Dov,

> an affluent man and a man of means, forty-five years old. He wore a short coat, and in general he was dressed honorably. Nevertheless, he would look with contempt and revulsion on all the fields of knowledge the Talmud did not encompass. His name was known among the learned, since he gained his knowledge in all fields of wisdom from *Sefer ha-Brit.* He believed with all his heart that all the wisdom of the world originated in the Talmud. When referring on occasion to one of the gentile scholars who excelled in a particular discipline of science, he would add: "Without a doubt, he knew the Talmud."[74]

Writing In roughly the same period and country, the enlightened orthodox rabbi and Hebrew essayist Samuel Alexandrov (1865–1941) had nothing but praise for Hurwitz because of his remarkable chapter on the love of neighbors: "And I declare, if *Sefer ha-Brit* had been created for only this honorable discourse *'ahavat re'im,'* it would have been enough."[75]

In a series of essays written by young eastern European Jews in the late 1930s and collected by the Yivo organization, yet another reference to Hurwitz is preserved. A young boy who called himself Henekh and was a student at the Ḥafeẓ Ḥayyim yeshiva in Radon wrote one of the essays in Yiddish, which includes the following: "I obtained a copy of *Sefer ha-Brit* . . . and virtually committed it to memory, reading it in the bathroom for fear of being caught and confronted with a whole new series of accusations. *Sefer ha-Brit* gave me a sound foundation in anatomy, physics, geography, and the like. I had a weakness, however, for showing off my scientific learning to my friends without telling them its source. This led to my becoming known as a person of wide-ranging knowledge, and I was sought after by those who were drawn to the *Haskalah* [Jewish enlightenment]."[76]

As this short inventory of literary references demonstrates, Hurwitz's book caught the imagination of many writers, especially those who reflected on their traditional Jewish upbringing, whether nostalgically or satirically. In such a context, *Sefer ha-Brit* held an honorable place as a work legitimately belonging to the tradition itself but also pointing in a direction beyond it, toward the wider horizons of Western culture. That creative tension of being both within and without was at the very root of the book's success.[77]

In closing this chapter, I offer two final references to *Sefer ha-Brit* of a somewhat different sort. In the course of time, the book attained a kind of mythical status, transforming the author into a cultural hero. A story that exists in at least two versions offers the following account of a man pretending to be the author of the book:

> The author of *Sefer ha-Brit* was Rabbi Pinḥas Eliyahu, may his memory be a blessing. When the book was first published, it appeared without the author's name. One time a deceitful person came before Rabbi Ẓevi Hirsch, the head of the rabbinical court of London, and bragged that he was the author of *Sefer ha-Brit*. But the rabbi knew he was a liar, since he knew the name of the actual author. He honored him by offering him an opportunity to present the novel ideas in the book, but the man had nothing to say. This rabbi then declared to him: "You are only the *sandak* [the one who holds the baby on his lap during the ritual circumcision, the *brit*] not the *ba'al ha-brit* [the father of the circumcised child; a play on the word *brit*, which refers both to circumcision and to the 'the book of the covenant']!"[78]

This version of the story is not entirely removed from the reality of the author and his book. The context of deception surely recalls Hurwitz's unpleasant experience of seeing a pirated version of his book. It is also not coincidental that the rabbi of the story is none other than Ẓevi Hirsch Levin, the rabbi Hurwitz actually met in Berlin and, according to Moses Koerner, who hosted Hurwitz in his house. Levin had also been the chief rabbi of London, thus giving this version of the story an English provenance. In the second version, the deceitful man is a bookseller who, having purchased hundreds of copies of the book to sell, presented himself as the author. This aspect of marketing the book also reflects Hurwitz's reality.[79]

The second story is of more recent origin and emerges in the circle of Rabbi Elazar Shach (1899–2001), an ultraorthodox rabbi and head of the Ponovitch yeshiva in Bnei Brak, Israel. Rabbi Mattisyahu Solomon, the spiritual guide (*mashgi'aḥ ruḥani*) of the Lakewood yeshiva in New Jersey, reported that

> he first saw this book [*Sefer ha-Brit*] on the desk of the honorable rabbi Shach, and when he [Solomon] began to look casually through the book, the rabbi told him: "This book shall never be removed from my desk." When the *mashgi'aḥ* asked him the reason for this, he showed him the following passage [in *Sefer ha-Brit*, pt. 1, sec. 9, chap. 16, on the length of exile and the hope for the coming of the messiah] and said of himself: "When I read his [Hurwitz's] words, it serves me as a support [*ḥizzuk*] for the correct approach to our being still in exile."[80]

Shach's attachment to the book and its eloquent messianic message is not as idiosyncratic as it may seem. At least one other rabbi, Jacob Tessler, actually copied the entire text singled out by the rabbi of Bnei Brak in his own book *Be-Aḥarit ha-Yamim* (At the End of Days), published in Leeds in 1942 and republished in 2000, an edition Schach might have noticed before he died. After praising Hurwitz for speaking about the coming of a real messiah—as opposed to an abstract notion of a messianic era—Tessler adds: "I make a covenant [*brit*] with this righteous man Pinḥas Eliyahu, the author of *Sefer ha-Brit*."[81] For Shach and Tessler, this text was Hurwitz's most lasting legacy, a spiritual message of comfort suggesting that although the messiah tarries, he is bound to come when Jews open their hearts to him with proper behavior and good deeds. Schach, like Ḥatam Sofer and other traditionalists before him, had legitimated and sanctified this "encyclopedia" of the sciences for his own followers. No other work of its kind could claim such an endorsement.

Epilogue

I BEGAN this book by constructing an imaginary dialogue, or more precisely a study in contrasts, between a Jewish metaphysician of the eighteenth century, Naphtali Ulman, and his younger contemporary, Pinḥas Hurwitz, to evoke a sense of the changing intellectual and spiritual sensibilities of the mid-eighteenth to the early nineteenth centuries. I would like to conclude by constructing another dialogue or conversation, this time based on the very recent work of three contemporary scholars of Jewish history and thought, Jonathan Garb, Eliyahu Stern, and Olga Litvak, who have approached aspects of modern Jewish culture from an original and revisionist point of view. Each in his or her own way offers fresh ways of thinking about Jewish culture in the modern era, and each of their approaches resonates directly with the portrait of Hurwitz I have tried to paint in the previous pages. The three represent only a small part of a larger scholarly explosion of work on the Kabbalah in the modern era, on the cultural dynamics of modern eastern European Jewish culture, and on the meaning and impact of the *Haskalah*, the so-called Jewish enlightenment, on the shaping of modern Jewish thinking.[1] In privileging these three scholars, I do not claim their work is without flaws or that they address the subjects they describe exhaustively. Their importance lies in the fact that they have raised new questions or offered nuanced perspectives on matters of relevance in appreciating *Sefer ha-Brit*, its popularity, and its special niche in the culture of modern Jewry.

All three scholars begin with the assumption that the dominant model for understanding the history of modern Jewish society proposed more than fifty years ago by the Israeli historian and sociologist Jacob Katz is in need of serious revision. They especially object to his position that the modern era represented a radical break from the past and that the primary transformation of modern Jewry was engendered by secularization, the rational assault on religion first raised by the Enlightenment, acculturation, and in some cases, radical assimilation.[2] For Jonathan Garb, secularization was only one path of modernization for Jews, who addressed the challenges of modernity in multiple ways, depending on the local conditions of their communities and on the cultural values and ideas they absorbed from their host environments. Focusing on the fate of the kabbalah in the modern era, especially in the circle of Moses Ḥayyim Luzzatto in Italy, Garb contends that another response to the modern world lay in mystical revitalization. Comparing the political and military metaphors of Luzzatto's kabbalistic discourse with the philosophical discourse of his Italian contemporary Giambattista Vico, Garb sees Luzzatto as a political religious thinker whose nationalistic rhetoric was unrelated to secularism but rather emerged in a mystical religious context. The writings of Luzzatto and his disciple Moses David Valle, for Garb, reveal an internal transformation of religious language, not its abandonment.[3]

Garb's interest in the mystical currents of Jewish modernity is part of a larger project of recent scholarship to understand the radical changes in the kabbalah in the early modern and modern eras, a project largely ignored in the foundational scholarship of Gershom Scholem and his students. Garb emphasizes especially two ways in which the kabbalah was transformed in the modern era: through the translation of kabbalistic ideas into exoteric terms accessible to wider circles of Jews reading popular kabbalistic works in print and incorporating kabbalistic customs and notions into their ordinary religious behavior; and through its psychologization, shifting from a primary focus on supernal mysteries to a growing concern for the fate of the individual and his or her soul. These trends have begun to be delineated in the work of a new generation of kabbalah scholars and Jewish cultural historians, most of them Israeli, such as David Sorotzkin, Maoz Kahana, Ḥaviva Pedaya, Boaz Huss, Shaul Maggid, and especially Roni Weinstein. Weinstein has recently published an ambitious book that attempts to trace the roots of modernity to Lurianic kabbalah, situating it within the broader cultural and spiritual trends of primarily Catholic Europe.[4]

The path taken by Pinḥas Hurwitz fits quite well with the findings of this new scholarship. Not only did this book by a kabbalistic writer have a modernizing impact, but even more significant is Hurwitz's indebtedness to the Italian commentators of Lurianic Kabbalah, including Luzzatto himself, Jacob Ergas, and Emanuel Ḥai Riki, with their emphasis on cultivating the soul, inculcating moral values, and appropriating motifs and metaphors prevalent in their larger cultural environment. As mentioned earlier, Hurwitz composed a commentary on Riki's *Mishnat Ḥakhamim*, an accessible summary of Lurianic kabbalah that impacted Ḥasidic thought, especially in its discussions of the human soul, as Jonathan Garb points out.[5]

Eliyahu Stern's new book on the Gaon Elijah of Vilna, one of the major figures of eastern European Jewish culture in the modern era, begins with a critique of Katz's notion that the demise of traditional Jewish society was engendered by the rise of rationalism and secularization, a notion that he argues comes from Katz's teachers Max Weber and Karl Mannheim. He rejects the dichotomy assumed by Katz's approach between tradition and modernity in the first place, calling it misleading to dismiss traditional Jews as nonparticipants or as mere opponents in the process of modernization.[6]

Stern's major point is that the key to understanding the Gaon as a leader and thinker of modern Jewry is to take into account the cultural environment of his native city, Vilna, where Jews represented the majority of inhabitants in the eighteenth century. Unlike Jewish leaders in western Europe, especially Moses Mendelssohn, Elijah exemplified the emboldened leader of a majority culture neither particularly concerned with nor threatened by ideas and institutions of other religions or ethnic groups. Stern further contends that this stance of Elijah of Vilna expressed in embryonic form the unique experiences of modern eastern European Jewry. Accordingly, the Gaon cannot be explained by Katz's understanding of Jewish modernity. As a traditionalist, he did not express hostility to modernity nor did he share the experiences of western European Jews regarding emancipation, religious reform, and acculturation. And ironically, living in Vilna and not in Berlin, he developed ideas more daring than those of Mendelssohn, since he never felt threatened by a hostile Protestant culture. Nor did he feel the need to articulate his views with a careful regard for the sensibilities of the majority Christian society.

According to Stern, modernity was not just a movement based on liberal philosophical principles; it was, rather, a "condition that restructured all aspects of European Jewish life," including Ḥasidism, the yeshiva, and Jewish communal structures. For Stern, reflecting his focus on eastern Europe, the primary condition of modernity lies in the paradigm shift from code to Talmudic commentary and from traditional community (*kehillah*) to the traditional academy of Jewish learning (*yeshiva*), both expressions of religious privatization and the restructuring of authority in rabbinic Judaism. He boldly adds that given their unique demography as a majority culture, "Eastern European Jews developed a sense of agency rooted in a deep and pervasive investment in Jewish life and ideas."[7]

In focusing on the differences between eastern European Jews and their western counterparts, and in arguing for a more dynamic cultural world in the East, one "rooted in a deep and pervasive investment in Jewish life and ideas," Stern follows a recent trend in scholarship to privilege Jewish communities of eastern Europe that were more rooted in tradition, more learned, and subsequently the more authentic than those emerging in Germany and the rest of western Europe.[8] He is also indebted to the particular formulation of Gershon Hundert, who described Jewish modernity as experienced by eastern European Jews as the conspicuous absence of a "beckoning bourgeoisie" and the relative lack of conformity on the part of Jews living in the East to western values and norms.[9] Stern adopts Hundert's formulation and then enhances it by arguing that without the pressure of the outside and the need to fit into the norms of a majority culture, Jews created a more dynamic and uninhibited indigenous cultural space based on the self-assurance associated with being a majority, not a vulnerable minority.

Stern's emphasis on the Gaon's cultural environment and the psychology of Jewish identity in Vilna as opposed to Berlin offers food for thought regarding Hurwitz's own background and attitudes. Hurwitz, too, allegedly stemmed from Vilna and spent many of his adult years in Cracow. He was a traditional eastern European Jew who wrote primarily in Yiddish and Hebrew; he privileged Jewish learning, especially that of the Kabbalah, over non-Jewish texts, even while he boldly espoused an appreciation of science and of non-Jews. Surely the Gaon's approval of Barukh Schick's translation of Euclid, mentioned by Stern, is reminiscent of Hurwitz's stance toward the study of the natural world as well, as I argued in chapter 2. I might add, following a suggestion of Elḥanan Reiner, that Hurwitz's radical views on Jewish-Christian relations also betray his Jewish background as shaped in Vilna and Cracow. Other eastern European Jewish thinkers, Reiner points out, unreservedly professed tolerant and even appreciative stances toward non-Jews, thinkers such as Nathan Neta Spira, the Gaon himself, Jonathan Eibeschütz, Joshua Heshel Ẓoref, the aforementioned Barukh Schick, Mendel Lefin, Naḥman of Bratslav, Menasseh Ilya, and Meir Simḥah ha-Cohen, to name only a few.[10] Has Stern thus captured the primary motivation for

Hurwitz's daring positions, namely his eastern European background, as prominently displayed also by the Gaon himself and his followers?

I am hesitant, however, to accept without reservation Stern's stark dichotomy between the East and the West, between Vilna and Berlin, as it might apply either to Hurwitz or even to Elijah of Vilna or Moses Mendelssohn. It seems too simplistic to propose that Jewish leaders from Vilna were dynamic and bold while those from Berlin were timid and felt threatened by their majority culture. I recall a similar comparison I once made between the political restraint of Mendelssohn in Berlin and the daring outspokenness of his contemporary David Levi, a Jewish leader and public intellectual of London.[11] If London was more conducive for Jewish temerity than Berlin, was Vilna less or more? Such comparisons render no clear-cut conclusion in the end. To attribute any cultural posture solely on the basis of one's place of residence, without considering other factors such as personality and psychological makeup, is problematic.

Was Hurwitz open to the sciences and to loving non-Jews because he had benefited from the security of living among the Jewish majority of Vilna? It seems more likely that Hurwitz's peripatetic life, traveling from country to country and from the East to the West and back to the East, facilitated multiple contacts and influences that shaped his life choices and his ideological positions. Hurwitz exemplifies a kind of intellectual figure that is very much a part of modern Jewish life: a transnational or transregional Jew. He was an eastern European Jew educated in the West who was stimulated by a variety of environments. In the end, that condition of mobility was surely more typical for many Jews living in the East than Stern's stationary model might allow.[12] As Olga Litvak proposes, Berlin at the end of the eighteenth century was an organic part not of western but rather of eastern Europe, and Hurwitz, in visiting the city, found many others who, attracted by its "beckoning bourgeoisie," had made their way there from Vilna and elsewhere in the East.[13]

In the previous chapter, I mentioned the entry of Samuel Joseph Fuenn on Pinḥas Hurwitz in his history of Vilna's Jewish intellectuals. Fuenn referred to the confusion over the authorship of *Sefer ha-Brit*, with some attributing it to the Gaon and others to Mendelssohn. One might infer from his remarks that Fuenn considered Hurwitz a combination of both men and that this westernized eastern European Jew could not easily be pigeonholed as a Jew stemming either from the East or from the West. He was, in fact, both.

Olga Litvak's recent book on the *Haskalah*, *The Romantic Movement in Judaism*, opens with the statement that the conventional connection made by modern Jewish historians between the *Haskalah* and the Enlightenment is not self-evident. By the time the first Maskilim had begun to have a discernible impact on Jewish thought, she argues, the underlying premises of Enlightenment rationalism were already being challenged. A crisis of "enlightened self-scrutiny" was well under way, associated with the writings of Jean-Jacques Rousseau and Immanuel Kant. In fact, it was these two thinkers who had the most profound impact on Jewish thought. If there was a Jewish enlightenment, therefore, it was not the *Haskalah* but an early modern phenomenon, in progress well before what we now label the *Haskalah*. To understand the genesis of the *Haskalah*, in other words, one needs to appreciate that it did not develop anywhere near the western European or North American epicenters of the Enlightenment but in eastern Europe, "a region where the diffusion of enlightenment ideas became associated with the expansion of state power, not with revolutionary politics, scientific discovery, or the decline of religious belief." Litvak adds that the *Haskalah* had no independent republic of letters, that Maskilim lived in the shadow of eastern European enlightened absolutism, and that their formative experience was not revolution but the Polish partitions of the end of the eighteenth century.[14]

As mentioned above, Litvak suggests that when the *Haskalah* began, Berlin was actually located in eastern Europe, well east of the dividing line between "Ashkenaz" and "Polin." Moreover, the first generation of Berlin Maskilim were newcomers to the city, mostly from eastern Europe. Thus, from its inception, the *Haskalah* was transnational and eastern European, typified by the members of Moses Mendelssohn's circle in Berlin.

For Litvak, then, the proper context of the *Haskalah* is not the Enlightenment but the romanticism that was prominent in the second half of the eighteenth and early nineteenth centuries. The romantics, in short, initiated the critical reception and imaginative revision of Enlightenment ideas. And the Maskilim were neither secular intellectuals nor rationalists but founders of a new romantic religion that criticized both the false enlightenment and the false piety of the Ḥasidic movement.[15]

In Litvak's portrait of the *Haskalah*, the aforementioned Moses Ḥayyim Luzzatto emerges as a heroic and inspiring force. Luzzatto introduced into modern Jewish thinking the notions of poetic inspiration and spiritual illumination. In linking the training of both the rational and the imaginative faculties in a process he called "*haskalah*," he advocated a form of saintliness that merged disciplined ethical behavior with spirituality and critical learning of the sciences (*ḥakirah*), all suffused in a kind of messianic aura.[16]

Of the three scholarly approaches I have discussed here, Litvak's interpretation surely correlates best with the portrait of Pinḥas Hurwitz presented in this book. He was an eastern European Jew who was transnational. He was deeply indebted to Kant and his critique of metaphysics. While he does not explicitly mention Rousseau, he had undoubtedly absorbed Rousseauian ideas, as evidenced by his discussion of natural society and the social contract in his chapter on the love of one's neighbors. Most relevant of all is the prominent place Litvak gives to Luzzatto in describing the romantic *Haskalah*. Her analysis is all the more interesting in that it meshes nicely with that of Jonathan Garb, of whom she appears to be unaware, thus suggesting that both scholars reached a similar conclusion independently of each other.

Hurwitz's intellectual product is perhaps best understood in the mold of Moses Ḥayyim Luzzatto, who likewise merged kabbalah and science, invested deeply in moral education and the pursuit of saintliness, and sought spiritual illumination and connection with the divine. It is possible that a messianic urge also informed his enterprise. At least some of his readers, as we saw in the last chapter, called this messianic dimension the most attractive feature of Hurwitz's book as well.

In the end, I will refrain from labeling Hurwitz either exclusively a Maskil, or a romantic, or an orthodox Jew. His multiple involvements and passions, coupled with his mastery of the book market, belie any particular ideological orientation or group affiliation. He was simultaneously many things: a Maskil, a kabbalist, a moral and spiritual practitioner, a student of the sciences critical of metaphysics, a social critic with radical ideas regarding Jewish-Christian relations, a consummate peripatetic crossing national and cultural boundaries, and, last but not least, a shrewd businessman. I offer this reconstruction of his intellectual biography as a modest but important narrative within the larger story of Jewish modernity. In juxtaposing the history of his life and legacy against some recent thinking about the shaping of modern Jewish culture, I hope I have made a stronger case for his enduring significance and relevance.

APPENDIX 1

Editions of *Sefer ha-Brit*

EDITIONS OF THE ENTIRE WORK

1. Brünn, 1797. The first, anonymous edition was published by Joseph Karl Neumanns and Joseph Rossmann in two parts. It contains censor pages in Yiddish and German signed by Karl Fischer on January 21, 1799, in Prague, followed by seven rabbinic *haskamot.* The beginning of the introduction contains the Hebrew phrase "*shiviti adonai lenegdi*" (I have set the Lord before me; Psalm 16:8), with the initials of the author. Part one with one introduction is 265 folios; part two is 137 folios. There is a list of mistakes on 131a. This version contains an additional text of Beer Oppenheim of Pressburg, but it is removed in all later editions.
2. Brünn, 1800–1801. This edition was published by Joseph Rossmann alone, apparently in two parts (although Hurwitz claimed in the second introduction to the later expanded edition of 1806–7 that only the first part of the full work appeared in this edition). This is the pirated edition, with rabbinic *haskamot* removed. It is not clear whether there were originally two versions, one published with the second part and one without (which might explain Hurwitz's claim). In the version I inspected at the National Library in Jerusalem, the title page refers to both parts of Hurwitz's composition. Part one contains 209 folios in twenty-one chapters. The second part, comprising 128 folios, has its own title page, mentioning that "most of its words follow the opinion of the teachings of the holy Isaac Luria in his book *Eẓ Ḥayyim* based on *Sha'arei Kedushah*, the book of his student Ḥayyim Vital."

   On the last page of the version I looked at, a paragraph appears indicating that the book was published several times, and that this version is larger than the first, with 390 additions, double the size of the original. This same paragraph appears in the Zolkiew edition of 1806–7, which mentions 350 additions. Since Hurwitz claimed in his second introduction (to the edition of 1806–7) that these additions were included only to make this pirated edition obsolete, it remains a puzzle how this paragraph appeared already in 1800–1801. Perhaps this was a later version of the pirated edition, published after 1807 and mixing the pages of one volume with the other.

3. Zolkiew, 1806–7. This edition was published by Abraham Judah Meir Hapfer in two parts. (On Hapfer, see Ḥayyim Friedberg, *Toledot ha-Defus ha-Ivri be-Polanya* [Antwerp: published by the author, 1932], 65–66; hereafter cited as Friedberg, *Polanya*.) Hapfer opened his print shop in 1793 and continued to publish until his death in 1811. This is the first time that the volume was called *Sefer ha-Brit ha-Shalem* (The Complete Book of the Covenant), and it served as the basis for all subsequent editions. On the title page, it is explained that the book is called "complete" because the author added 350 additions to the original text and the book has two introductions that should be read in their entirety. This Zolkiew edition, it is claimed, is published with bigger and nicer letters than the original edition published in Brünn. Unlike the previous editions, the author's name now appears. On the last page, 185b, the same paragraph mentioning the additions appears as in the 1800–1801 edition that I inspected in Jerusalem, but 350 are offered, not 390. Part one ends on page 122a; part two runs from 122b to 185b. A list of corrections then appears, followed by instructions on how to purchase the book: the author is selling his book in Cracow where he is residing, and "those who have signed up for the book in the state of Hungary can search for it in the city of Pest on market day . . . and in the state of Poland, they can search for it in the city of Lemberg."
4. Vilna, 1817–18. This edition was published by Menaḥem Man ben Barukh, also called Menaḥem Man Romm (on whom, see Friedberg, *Polanya*, 88). The censor page follows the introduction to part one. On page 4 a notice states that use of the term *goy* does not refer to present-day Christians. This paragraph seems to

serve political ends and appears superfluous in the light of Hurwitz's chapter on loving one's neighbors. (A similar statement also appears in the Zolkiew edition on the inside cover page.) Part one consists of 191 folios; part two is 104 folios. On the last page, the same paragraph mentioning that the book is twice as large with 350 additions appears. The name of typesetter and corrector, Eliezer Yeruḥam ben Yehoshua of Vilna, is also mentioned.

5. Salonika, 1846–47. This is a Ladino translation, published by Shabbetai Alaluf and Yiẓḥak Jahon and translated by Ḥayyim Avraham Benvenisti Gateňo. Part one consists of 178 folios; part two, 198 folios.

   Elena Romero, *La creación literaria en lengua sefardí* (Madrid: Editorial MAPFRE, 1992), 135–36, claims that this text reappeared in later Ladino versions of 1881 and 1900. She is not sure to what extent the version published in Constantinople in 1900 by Eliyahu Levi ben Naḥmias is based on this one. She is only familiar with the first volume of this translation, which includes the first ten chapters of part one of *Sefer ha-Brit*. Cf. Matthias B. Lehmann, *Ladino Rabbinic Literature and Ottoman Sephardic Culture* (Bloomington: Indiana University Press, 2005), 7. Dr. Katja Smid is currently working on this and the other Ladino versions of *Sefer ha-Brit*.

6. Lemberg [Lvov], 1859. This edition was published by Shrenzel—not Asher Lemel ben Meir David Shrenzel, who ceased publishing in 1820, but his son David. (On Asher, see Friedberg, *Polanya*, 83–84.) David Ẓevi Hirsh Shrenzel supervised the press of Feiga Grossman until 1858–59 and then went out on his own with other partners (ibid., 95).

7. Lemberg [Lvov], 1865. This edition was published by Michael Franz Poremba in two parts (140 + 74 pages), lacking titles and subtitles (see Friedberg, *Polanya*, 85). Poremba opened a general press in 1850, with a special section for Hebrew books that lasted until 1879.

8. Warsaw, 1869. This edition was published by Isaac Goldmann (1812–87; on whom, see Kenneth Moss, "Printing and Publishing: Printing and Publishing after 1800," in the online *Yivo Encyclopedia of Jews in Eastern Europe*, www.yivoencyclopedia.org/article.aspx/Printing_and_Publishing/Printing_and_Publish

ing_after_1800). Goldmann ran his own press from 1867, producing more than one hundred books, among them Talmud tractates. (See also Friedberg, *Polanya,* 86–87, 111–12; and Isaac Kornfeld-Dagani, "On the History of Hebrew Printing in Warsaw from the Beginning of Printing in 1796 until the Holocaust in 1939" [Hebrew], in *Perakim* [Jerusalem, 1968], 353–55.)

9. Warsaw, 1870. This is the Isaac Goldmann edition consisting of 190 folios (380 pages), published by Jacob Joseph Kalinberg of Warsaw and Jacob Eliezer Edelstein of Bialestock (both deceased at the time of publication!).
10. Warsaw, 1871–72. This edition was published by Yoel Lebenzohn, 191 folios. In 1852, Avigdor Lebenzohn left for Israel and gave the press to his son, Yoel. Yoel died in 1872. He and his sons published the journal *Ha-Zefirah.* (See Kornfeld-Dagani, "On the History of Hebrew Printing," 350.)
11. Warsaw, 1873. Published by Isaac Goldmann, 192 folios.
12. Warsaw, 1876. Published by Isaac Goldmann, 360 pages.
13. Warsaw, 1879. Published by Isaac Goldmann, 360 pages.
14. Warsaw, 1880. Published by Isaac Goldmann, 360 pages.
15. Warsaw, 1881. Published by Isaac Goldmann, 360 pages.
16. Salonika, 1880–81. This Ladino translation was published by Raphael Yiẓḥak Benvenisti. The title page reads, *Sefer ha-berit: berakhah ha-meshuleshet o las tres luzes . . . el kual avla de los siete sielos, de los diez gilgulim . . . tradusido por Refael Yitshak Benvenisti.* Salonika: Estamparia de Hevrat Ets ha-Ḥayim, 641 (1880 or 1881). It consists of 178 pages. According to Romero (see above, no. 5), this was republished in 1900. More recently, Katja Smid has reexamined the 1880–81 edition and discovered two versions. The first version of 88 leaves alone was part of an anthology of four books: *Sefer ha-Brit* (1–88a), *El rijo de la vida (Orekh Ḥayyim?)* (1–60b), *Ba'al teshuva* (1–39b), and *El asolado en la isla* (40a–88a), the latter being a Ladino version of *Robinson Crusoe.* The other two works were apparently moralistic/homiletical works, not yet fully examined by any researcher. The second version was published alone in 178 pages, as Romero reported.
17. Warsaw, 1887. This edition of 360 pages was published by Isaac Goldmann, with his name explicitly mentioned. The title page claims that the book includes all the sciences of the world and

ḥiddushei Torah, musar haskel v ḥokhmat ha-emet (new interpretations of the Torah, enlightened morality, and the wisdom of truth [kabbalah]). On the second title page one reads: "One who buys this book buys a culture of books from the scholars of the world, and one who reads it two or three times will satisfy his soul with all these sciences." The first part represents only an introduction to the second, which is the essence of book. In the second part, one finds the treatise on loving one's neighbors, which the author claims is totally unique, for "one who reads this acquires an entire library and a teacher for himself."

18. Warsaw, 1889. Published by Isaac Goldmann without his name in 360 pages.
19. Warsaw, 1893. Published by F. Razin and B. Kalinberg at the press of Doverosh Tursh. It contains 360 pages and is identical to the edition of 1889.
20. Vilna, 1896–97. Published by Abraham Rozenkranz and Menaḥem Schriftsetzer in 80 pages. (See Friedberg, *Polanya*, 131, on this press established in 1863.)
21. Warsaw, 1898. This is a Yiddish translation by Yosef Meir ben Shemuel Yitshak Yavetz (1832–1914), printed by Doverosh Tursh, in 160 pages. (For more about this press, which was founded in 1886, see Kornfeld-Dagani, "On the History of Hebrew Printing," 359.) It is printed as 104 chapters but lacks a two-part division and bears no clear relationship to the Hebrew original. On the last page of every edition published in 1898 there is a bill of sale "from Yavetz to Tursh" in Hebrew.
22. Constantinople, 1899–1900. This is a Ladino translation of the Arditi Press published by Eliyahu Levi Ben Nahmias in 178 pages. Romero is not sure this is related to the Salonika edition, but see Lehmann, *Ladino Rabbinic Literature*, 239. Katja Smid claims that this edition of *Sefer ha-Brit* (3–178) was published with two other books: *Ba'al teshuva* (3–88), and *El asolado en la isla* (3–146). *El rijo de la vida* (see no. 16 above) was apparently omitted.
23. Vilna, 1899–1900. Published by Yiẓḥak Funk from the Goldmann edition in 360 pages.
24. Petrikov [Pietrokov], 1904. Published by Mordechai Tsederboym in 392 pages. (On his press, established after 1900, see Friedberg, *Polanya*, 165.)

25. Vilna, 1904. This edition was edited by Yiẓḥak Funk and published by Abraham Ẓevi Rozenkranz and his brother Menaḥem Schriftsetzer, based on the Warsaw edition of 1889, in 360 pages.
26. Vilna, 1911. This was edited by Yiẓḥak Funk, based on the Vilna edition of 1899–1900, and contains 260 pages.
27. Vilna, 1912–13. Published by Yiẓḥak Funk; based on the Vilna edition of 1899–1900.
28. Petrikov-Warsaw, 1913. Edited by Mordechai Tsederboym. It mentions on the title page that it is from the shop of Abraham Kahana in Warsaw. It contains 392 pages.
29. Vilna, 1919–20. Edited by Abraham Rozenkranz and Menaḥem Schriftsetzer; based on the Petrikov-Warsaw edition of 1913.
30. Warsaw, 1919–20. Published by Abraham Kahana; based on the Petrikov-Warsaw edition of 1913.
31. Warsaw, 1928–29. This is a Yiddish translation edited by Avraham Yosef Klaiman and published by Y. Vagmeister in 160 pages. This translation was in a volume entitled *Maẓmiaḥ Yeshuot,* edited by a certain Mendel Ravitzki and including a long list of *haskamot* by several Ḥasidic rabbis (Sadigura, Viznitz, Munkatch, and others). This was clearly a Ḥasidic book, and Ravitzki obviously considered *Sefer ha-Brit* appropriate for the volume. My thanks to Professor Yitzhak Melamed for this information.
32. Bnei Brak and Jerusalem, 1959–60. Based on the Petrikov-Warsaw edition of 1913, printed in 392 pages. It contains a new title page stating the city of publication as Jerusalem, but on the inside Bnei Brak is mentioned.
33. Brooklyn, 1969. A Yiddish translation of 160 pages, 104 chapters, translated/co-authored by Yosef Meir ben Shemuel Yavetz (1834–1914), and based on the Warsaw edition of 1898.
34. Brooklyn and Bnei Brak, 1977. A reprint copy of Petrikov-Warsaw 1913, consisting of 392 pages published by Moshe Menaḥem Blum, with a handsome embossed cover.
35. Jerusalem, 1980–81. A copy of the Petrikov-Warsaw edition of 1913.
36. Jerusalem, 1989–90. This attractive edition is called *Sefer ha-Brit ha-Shalem,* printed in larger letters and in a corrected version of the text of 710 pages, including the restoration of censored portions of previous editions by S. Krauss, with a new introduction.

The volume has a shiny leather cover on which a globe is at the center, highlighting the centrality of the Land of Israel.

PRINTED EXCERPTS OF THE WORK

37. Salonika, 1842–43. This is the separate publication of Hurwitz's discussion of loving one's neighbor, translated into Ladino with the title *Sefer Darkei ha-Adam: Il rov de sus palabras son de el sefer ha-berit helek sheni, ma'amar 13* "Ahavat Re'im." Yitshak Bekhor Amarachi and Yosef ben Meir Sasson were the translators and Sa'adi ha-Levi was the publisher who printed the text on 92 leaves. (On the two translators, see Romero, *La creación literaria*, 113.) The translators added excerpts from *Seder ha-Dorot* of Yeḥiel Heilprin and from *Shevet Yehudah* of Solomon Ibn Verga explaining in the introduction: "And we added a few small things of our own, and there are spiritual things that ease the anxiety from our hearts and the sadness of knowing what happened to us during the time of Spain."

    See also Lehmann on these two collaborators, in *Ladino Rabbinic Literature*, 92, 150–51, 177–82, 189–93; and Yiẓḥak Emanuel, "Printing Houses and Printers" (Hebrew), in *Zichron Saloniki*, ed. David Recanati (Tel Aviv, 1986), 2:242. These two men also published *Musar Haskel* (Salonika, 1842–43, 1848-49, 1889–90), which appears to be heavily indebted to *Sefer ha-Brit* on a variety of scientific matters, including smallpox. This work is discussed extensively by Lehmann and by Romero.
38. Salonika, 1848–49. This edition, apparently the same as 37, was published by Daniel Frag'i in 55+ leaves; it was subsequently republished in 1900.
39. Vienna, 1810. Partial edition by Anton Schmid of 128 leaves of only part two.
40. Salonika, 1900. Reprint of *Sefer Darkei ha-Adam*, Salonika edition of 1842–43; same as nos. 37 and 38.

APPENDIX 2

Hurwitz's Instructions on Printing His Book, from His Second Introduction

SEFER HA-BRIT HA-SHALEM (JERUSALEM, 1990), 21–23:

This is my covenant [*brit*] in printing this second edition with additions. I do not come to extend and lengthen the duration of the prohibition on publication for any more years. Thus beginning in 1809 and subsequently, it is permitted for anyone to publish it. This is my desire in order to increase the Torah and to enhance wisdom so that intelligence will grow among my people. But this will only be satisfied by the publisher if he will fulfill the following conditions, which I enumerate below:

1. He will publish this later edition, not the first edition, since this is complete unlike the first.
2. The paper will be good and attractive but strong.
3. The letters will be large or mid-size but not small.
4. One should not add nor subtract anything from the book, neither from the first nor from the second introductions, not anything, nor half of anything.
5. The end of the first part [of the book] should be published on the same page with the beginning of the second part. The end of part one should conclude in the first column and the second part should begin on the second column of the same page so that they will be united and not separated. One part should not be separated from the other, nor the first part from either the second or

the second from the first. There is no part that can stand on its own, as in the case of the parts of other books. Each part of this book is not like the others, since I carefully preserved the order of God's redemption in it and it is connected from the beginning of the first part until the end of the second; every discourse is fastened to the preceding one and to the one next to it, and similarly each chapter. And thus all the words of the second part are dependent and rest on the words of the first part and without part one, enlightened wise men will not comprehend anything. The first part is only an introduction to the second part and it alone represents a gate to the house. A person who possesses only the first part is a gatekeeper, not a house owner. So in what way is it possible to divide this book? Therefore [the two parts] will be attached together on one page and the connection will be strong so that they will be stitched together as a permanent possession in order that the people will stand in a perfect covenant.

6. There should not be an index of any chapter or discourse regarding any specific matter or another or to a specific page about any other matter as is the custom in other books, where an index exists to find a desired item at the book's beginning or end. One should not follow this practice in this book for two reasons: First, because most of the matters in this book are not concentrated in one place alone but are mentioned in many places, scattered and separated in details in many discourses and chapters. One detail is mentioned in one place and another elsewhere and they are not contiguous to each other. Thus they will be inaccessible to a reader who will be unable to locate them through a table of contents nor will he find an index useful to him. Even the titles of discourses and parts refer only generally to the matter in this book. Moreover, with all this, one may find many words of Torah and fear of Heaven in the first part devoted to human wisdom and much human wisdom in the second part devoted to Torah wisdom and divine knowledge. So quite often one will discover a subject deemed generally unrelated to a specific section in the section itself. In sum, her course meanders [Proverbs 5:6] so that you cannot know several matters completely from one location. Accordingly, one cannot find his desired subject from one hundred indices. Even one who reads and studies several discourses

and chapters of this book will not be able to understand any matter completely until he reads the entire book from cover to cover in order and without skipping anything. I have already mentioned in the beginning of the first introduction why I failed to collect and record all the details and parts in one place and why the order of our recollections in this book is very good and even marvelous. It is because of how all the subjects are bound together in it and fit into one general composition called harmonic. Lift your eyes up and behold from this location and you will know it. The second reason why I have not included an index is hidden to me and it is also because of an important reason. So don't alter my instructions.

7. The format of the book should be quarto, which is four leaves called *Bogen Papier* [full unbound sheets of paper] and not half a *Bogen*. However, to print it as octavo is permitted and this is a good and attractive thing.
8. Abbreviations should not be seen in the book except for those already in the text. One should never introduce any abbreviations, even those well-known ones such as א"כ, which everyone knows means אם כן. One should be careful never to print one letter instead of a complete word or a second letter in place of a second word or two and so on. Any member of the Jewish community who will fulfill the words of the covenant regarding the printing press and will remove any abbreviations that are already in this book and will sacrifice perfect "burnt-offerings" by making all the words complete in the book of the covenant will certainly know that God will pay his reward and the work of his hands will be desired. Almighty God will bless him from that day on and God's blessing will be bestowed to everything he possesses. Everything he does will be rewarded by God.
9. There should not be a word missing a final letter in the entire book. One is forbidden in this book to draw a meaningless horizontal line [*kav tohu*] above the word indicating that the last letter is missing, as is the custom in other books. The printer or the publisher should be careful [to supervise] the typesetter, who knows how to join the letters from iron and lead. The latter is called a *Setzer* [compositor, typesetter], who attaches all the letters to the words so that all of them will be perfect. It is occasionally his cus-

tom to remove at will the last letter of the last word to make the work easier so as not to spoil the line but to align it a second time in an appropriate manner. For their sinews are full of fever [Psalm 38:8] for this manner has brought shame to every "smelter" who destroys and soils the holy books, the divine words of eternal life because in the heart of every Jew is the love of perfection. One should also supervise the *Setzer* to make the ends of lines perfectly straight, since it is often the custom of the typesetter who does as he pleases to fill out an empty space at the end of a line with an additional letter to straighten the line, filling the empty space with a letter that will begin the next line below it. The traitor comes in place of the upright [Proverbs 21:18]. The way of a fool is right in his own eyes [Proverbs 12:15]. He straightens in an evil way, and all this is not worthwhile.

And you who cling [a play on the word typesetter, *medavek,* which also means one who clings (to God)] to the Lord your God, I swear to you, if your hearts are aroused to add even one abbreviation or to leave off the last letter of a word in this book and to put in its place a line above to indicate the missing letter or to fill the empty space at the end of a line with a letter that begins the next line, please be careful with such things in this book. Rather set all the letters of each word and attach them [a play on the root *zaraf,* meaning both "attach" and "refine"] as one who refines silver, making the ends of the lines perfect and straight, fixing the horizon [Proverbs 8:27] with a straight ruler upon the portion of the printer [Deuteronomy 33:21] that is a square wooden board [the form locked in the frame] upon which iron and lead letters are attached. Be careful to attach the letters on the board so that all of them are done correctly.

10. I have seen the stress, that is, the pressure applied in the printing house, the craft called *Pressen Zieher* [pulling the lever to press the type into the paper]. I am disgusted with my life [Genesis 27:46] because of their pressers and pushers, most of whom are slothful and will not press with full human strength [so as to make a good, solid impression on the paper with ink equally uniform across the sheet]. Subsequently the letters are not properly absorbed with ink on the book, especially if the paper is fit for writing called *Schreibpapier;* these letters will not be well recognizable because they will

be filled with bright white spots that turn to white. Thus the reader can only make out what is marked, while the ink will only reach the border of its destination on the paper as "ghosts" [*refa'im*, that is, weak impressions]. Consequently, the printer should carefully supervise his pressers and he should order them to push and press hard the printing press with a strong hand so that the paper can fully absorb the letters on the wooden board. With courage and great force one should fortify weak hands.

11. One should carefully supervise the proofreader, who will do his holy work with good concentration free of wandering thoughts and should forbid anyone to speak with him regarding essential matters, business affairs, or even the affairs of "kings" at the time he is proofreading. Woe to the chirpers and moaners [Isaiah 8:19]. Woe and desolation is destined for the proofreader who reads falsely and acknowledges a mistake.
12. As long as I am living on this earth, no one should make a shortened version of this book or a section of it. He should never produce a single discourse or part of it to stand alone. One who shortens it will shorten his life, and one who divides it will divide his years [of life]. The most important point is that it should be proofread very well.

APPENDIX 3

The Contents of *Sefer ha-Brit*

(Page numbers follow Jerusalem edition, 1990)

Part One: *Ketav Yosher* (A Righteous Composition), 4–395[1]
First Preface, 4–16
Second Preface, 16–24
Introduction, 24–32
Chapters 1–4: Cosmology, Astronomy
Chapters 5–9: The Elements, Geography
Chapter 10: Meteorology
Chapter 11: Minerals, Plants, Animals, Man in general
Chapter 12: Mineralogy
Chapter 13: Botany
Chapter 14: Zoology
Chapters 15–16: Psychology (soul of plants and animals)
Chapters 17–21: Man (Embryology, Anatomy, Psychology, the Intellect and Man's Knowledge)

Part Two: *Divre Emet* (Words of Truth), 396–610
Chapter 1: The Souls of the Israelite
Chapter 2: Good and Bad Attributes
Chapter 3: The Divine Commandments
Chapter 4: Love and Fear
Chapter 5: The Righteous and the Pious Person
Chapters 6–10: Prophecy
Chapters 11–12: The Holy Spirit
Chapter 13: Love of One's Neighbor
Chapter 14: Love and Joy

Notes

PREFACE

1 Frances A. Yates, *Giordano Bruno and the Hermetic Tradition* (Chicago: University of Chicago Press, 1964); idem, "The Hermetic Tradition in Renaissance Science," in *Collected Essays: Ideas and Ideals in the North European Renaissance* (London: Routledge & Kegan Paul, 1984), 3:227–46.

2 David Ruderman, *Kabbalah, Magic, and Science: The Cultural Universe of a Sixteenth-Century Jewish Physician* (Cambridge, Mass.: Harvard University Press, 1988).

3 Several chapters of my *Jewish Thought and Scientific Discovery in Early Modern Europe* (New Haven: Yale University Press, 1995; Detroit: Wayne State University Press, 2001) are also relevant to the topic of kabbalah and science.

4 A few recent works that exemplify these trends are Eliyahu Stern, *The Genius: Elijah of Vilna and the Making of Modern Judaism* (New Haven: Yale University Press, 2013); Olga Litvak, *Haskalah: The Romantic Movement in Judaism* (New Brunswick, N.J.: Rutgers University Press, 2012); Shmuel Feiner, *The Jewish Enlightenment* (Philadelphia: University of Pennsylvania Press, 2003); Glenn Dynner, *Men of Silk: The Hasidic Conquest of Polish Jewish Society* (Oxford: Oxford University Press, 2006); Pawel Maciejko, *The Mixed Multitude: Jacob Frank and the Frankist Movement, 1755–1816* (Philadelphia: University of Pennsylvania Press, 2011); Maoz Kahana, "Bein Prague le-Pressburg: Ketivah Hilkhatit be-Olam Mishtaneh me *ha-Nodah be-Yehudah* ad *ha-Ḥatam Sofer*, 1730–1839," Ph.D. diss., Hebrew University, 2010 (soon to be published in an updated version). I refer to the works of Stern and Litvak in the epilogue of this book.

CHAPTER 1. THE HAGUE DIALOGUES

1 Previous scholarship on Hurwitz includes Ira Robinson, "Kabbalah and Science in *Sefer ha-Berit:* A Modernization Strategy for Orthodox Jews," *Modern Judaism* 9 (1989): 275–88; Noah Rosenblum, "The First Hebrew Encyclopedia:

Its Author and Development" (Hebrew), *Proceedings of the American Academy for Jewish Research* 55 (1988): 15–65; David Ruderman, "Some Jewish Responses to Smallpox Prevention in the Late Eighteenth and Early Nineteenth Centuries: A New Perspective on the Modernization of European Jewry," *Aleph* 2 (2002): 111–44; Resianne Fontaine, "Natural Science in *Sefer ha-Berit:* Pinchas Hurwitz on Animals and Meteorological Phenomena," in *Sepharad in Ashkenaz: Medieval Knowledge and Eighteenth-Century Enlightened Jewish Discourse,* ed. Resianne Fontaine, Andrea Schatz, and Irene Zweip (Amsterdam: Koninklijke Nederlandse Akademie van Wetenschappen, 2007), 157–81; Resianne Fontaine, "Love of One's Neighbour in Pinḥas Hurwitz's *Sefer ha-Berit,*" in *Studies in Hebrew Literature and Jewish Culture, Presented to Albert van der Hade on the Occasion of his Sixty-fifth Birthday,* ed. Martin F. J. Baaesten and Reinier Munk, Amsterdam Studies in Jewish Thought 12 (Dordrecht: Springer, 2007), 271–95; and most recently, Jeremy Brown, *New Heavens and a New Earth: The Jewish Reception of Copernican Thought* (Oxford: Oxford University Press, 2013), 133–43. Additional references to Hurwitz and his book are presented in subsequent chapters.

2 On Ulman, see Alexander Even-Chen, "Haskalah, Pragmatizim, ve-Emunah: Mishnato ha-Pilosofit shel Naphtali Herz Ulman," Ph.D. diss., Hebrew University, 1992; Zvi Malachi, "N. H. Ulman, Maskil and Philosopher" (Hebrew), *Studies on the History of Dutch Jewry* 2 (1979): 77–88; Shalom Rosenberg and Alexander Even-Chen, "Philosophical Letters at the End of the Eighteenth Century: Naphtali Herz Ulman and Moses Mendelssohn" (Hebrew), *Iyyun* 43 (1994): 209–20; Fred Van Lieburg, "On Naphtali Herz Ulman's Biography and the Reception of His Works in the Netherlands," *Zutot* 3 (2003): 58–65; Alexander Even-Chen, "Ulman Herz (Hartog Ulman) (1731–87)," in *The Dictionary of Seventeenth- and Eighteenth-Century Dutch Philosophers,* ed. W. van Bunge (Bristol: Thoemmes Press, 2003), 2:1003–6; and Reinier Munk, "Naftali Herz Ulman, een vroege maskil," Inaugural address, University of Leiden, November 4, 2003.

3 Based on a letter Hurwitz wrote to the Prague censor Karl Fischer, he may have known German, despite his explicit denial in *Sefer ha-Brit.* See my discussion in chapter 2.

4 He was buried on June 22, 1787, having died of a stroke. See Van Lieburg, "On Naphtali Herz Ulman's Biography," 61.

5 See Mezerich's *haskamah* at the beginning of *Sefer ha-Brit* and my fuller discussion of the chronology of Hurwitz's life in the next chapter.

6 See Irene Zweip, "Jewsh Enlightenment Reconsidered: The Dutch Eighteenth Century," in Fontaine, Schatz, and Zweip, eds., *Sepharad in Ashkenaz,* 297.

7 Naphtali Herz Ulman, *Sefer Ḥokhmat ha-Shorashim* (The Hague, 1781), 1a. Alexander Even-Chen translated this passage in his doctoral thesis, "Haskalah, Pragmatizim, ve-Emunah," English abstract, 11–12. I have tried to improve on his translation. The passage is also discussed by Shmuel Feiner in *The Jewish Enlightenment* (Philadelphia: University of Pennsylvania Press, 2003), 71–73.

8 *Sefer Ḥokhmat ha-Shorashim,* title page.

9 Hurwitz, *Sefer ha-Brit ha-Shalem* (Jerusalem, 1989–90), 4–7 (arabic numerals).

On Loewenstamm, see Zweip, "Jewish Enlightenment Reconsidered," 293, 300. On Azevedo, see Jozeph Michman, *History of Dutch Jewry during the Emancipation Period, 1787–1815: Gothic Turrets on a Corinthian Building* (Amsterdam: Amsterdam University Press, 1995), 125; and on Breslau, ibid., 153. On Mezerich, see Stefan Litt, *Pinkas, Kahal, and the Mediene: The Records of Dutch Ashkenazi Communities as Historical Sources* (Leiden: Brill, 2008), 124. On R. Isaac Ha-Levi of Lemberg, rabbi of Cracow, and his family connection with the rabbi of Rotterdam, see Majer Balaban, *Toldot ha-Yehudim be-Krakov uve-Kaz'imyez, 1304–1868* (Jerusalem: Magnes Press, 2002), 2:807.

10 Hurwitz discusses the whole sordid affair of the pirated edition (1800–1801) of the publisher Joseph Rossmann and his plan to revise and expand the book in the introduction to the second edition, *Sefer ha-Brit ha-Shalem*, 18–20. This introduction first appeared in the Zokiew edition of 1806–7 of Abraham Judah Meir Hapfer. This is all discussed more thoroughly in the next chapter.

11 See *Sefer ha-Brit ha-Shalem*, 555.

12 Ulman, *Sefer Ḥokhmat ha-Shorashim*, title page.

13 Immanuel Kant, *Kritik der reinen Vernunft* (Riga, 1781).

14 For a philosophical study of Ulman's work in relation to the philosophies of Leibniz and Wolff, see Even-Chen, "Haskalah, Pragmatizim, ve-Emunah." Succinct summaries with up-to-date bibliographies of the two German philosophers can be found in the online *Stanford Encyclopedia of Philosophy*, edited by Edward N. Zalta (http://plato.stanford.edu).

15 Ulman, *Sefer Ḥokhmat ha-Shorashim*, 2a–17b.

16 Ibid., 1b.

17 *Sefer ha-Brit ha-Shalem*, 324. The context of this comment is discussed further in chapter 4.

18 Ibid., 330. On the powerful tradition in Jewish thought that Hurwitz was emphatically undermining, see Abraham Melamed, *Rakaḥot ve-Tabaḥot: Ha-Mitus al Mekor ha-Ḥokhmot* (Jerusalem: Haifa University Press, 2010), and my discussion in chapter 4.

19 *Sefer ha-Brit ha-Shalem*, 344. This is also more thoroughly discussed in chapters 3 and 4.

20 Ibid., 360–61.

21 Ibid., 362–63. Hurwitz cited the opening of Solomon Maimon's *Givat ha-Moreh* (1st ed. Berlin, 1791; Jerusalem: Israel Academy of Sciences, 1965), 18. For a fine summary of Maimon's philosophy with up-to-date bibliographies, see Peter Thielke and Yitzhak Melamed, "Salomon Maimon," *Stanford Encyclopedia of Philosophy* (http://plato.stanford.edu). Also see chapter 4.

22 Naphtali Herz Ulman, *Ma'amar Selah ha-Maḥloket*, Ms. Leiden Hebrew 86/2, pt. 2, fol. 146a.

23 Ibid., fols. 147b–153b.

24 These themes are developed in Ulman's *Ḥokhmat ha-Olam*, Ms. Leiden Hebrew 87a, esp. fols. 143a–147b.

25 Naphtali Herz Ulman, *Ḥokhmat ha-Elohut*, Ms. Leiden Hebrew 87d, pt. 2, fol.

105a. The Bayle-Leibniz debate that took place between 1695 and 1699 is referred to on fol. 70a. On this controversy, see, e.g., David F. Norton, "Leibniz and Bayle: Manicheism and Dialectic," *Journal of the History of* Philosophy 2 (1964): 23–36; and Paul Lodge and Benjamin Crowe, "Leibniz, Bayle, and Locke on Faith and Reason," *American Catholic Philosophical Quarterly* 76 (2002): 575–600.

26 *Sefer ha-Brit ha-Shalem*, 189. I discuss this passage along with others in chapter 3.

27 Ibid., 463.

28 Naphtali Herz Ulman, *Ma'amar ha-Yiḥud*, Ms. Leiden Hebrew 88, fol. 4a.

29 Ibid., fols. 4a–6b; *Ḥokhmat ha-Nefesh*, pt. 3, Ms. Leiden Hebrew 126, fols. 5a–5b. On Ulman's Dutch book, see Van Lieburg, "On Naphtali Herz Ulman's Biography."

30 Naphtali Herz Ulman, *Ma'aseh ha-Tartuffe be-Lashon Ivrit*, Ms. New York JTSA 10169. This short pamphlet was copied in The Hague in 1776 by Israel ben Samuel Falk after his hearing of this sermon by Ulman. My thanks to Professor Reinier Munk for supplying me with a copy of the manuscript.

31 See, for example, *Sefer ha-Brit ha-Shalem*, 52, 358, and 360.

32 See, for example, ibid., 76, 375, and 520.

33 Ibid., 522–24, 537–45, and 569. See also chapters 2 and 5.

CHAPTER 2. PINḤAS ELIJAH BEN MEIR HURWITZ

1 On his date of birth, see, for example, Ben Zion Dinur, *Bemifne Ha-Dorot* (Jerusalem: Mosad Bialik, 1955), 265; Raphael Mahler, *Divre Yemei Yisrael: Dorot Aḥaronim* (Merhavia: Sifriyat Ha-Po'alim, 1956), 4:45; and Noah Rosenblum, "The First Hebrew Encyclopedia: Its Author and Development" (Hebrew), *Proceedings of the American Academy for Jewish Research* 55 (1988): 16, esp. n. 6. Rosenblum's reconstruction of Hurwitz's life is by far the best to date, and I have relied on it extensively, while supplementing, revising, and focusing on other aspects of Hurwitz's career not fully treated by Rosenblum. Hurwitz identifies Vilna as the city of his birth in *Sefer ha-Brit ha-Shalem*, 72 (where he also mentions the two lunar eclipses) and 382. The solar eclipse is mentioned on 73.

2 The citation is from Nathan M. Gelber, *History of the Jews in Buczacz*, Part 4, translated by Adam Prager, www.jewishgen.org/yizkor/buchach/buc045.html.

3 Gershom Bader, *Medina ve-Ḥakhameha: Toledot Kol ha-Ḥakhamim ve-ha-Sofrim she-Ariẓatam Amdah be-Galiẓia* (New York, 1934), 80–81.

4 Samuel Joseph Fuenn, *Kiryah Ne'emanah* (Vilna, 1860), 202–4. See also my remarks on Fuenn in chapter 6 below. Israel Klausner, *Vilna bi-Tekufat ha-Gaon* (Jerusalem: Reuven Maas, 1941), 46n.1, did not include Hurwitz among the writers of Vilna because he was uncertain he lived there. See also the newly edited version of Hillel Noah Maggid Steinschneider, *Ir Vilna*, ed. Mordechai Zalkin (Jerusalem: Magnes Press, 2002), where Hurwitz is also not mentioned.

5 See Eliyahu Stern, *The Genius: Elijah of Vilna and the Making of Modern Judaism* (New Haven: Yale University Press, 2013), with updated bibliography.

6 *Sefer ha-Brit ha-Shalem*, 20.

7 On Schick and his relation to the Gaon's circle, see David E. Fishman, *Russia's First Modern Jews: The Jews of Shklov* (New York: New York University Press, 1995), 22–45; and Yisrael Shapira, "Differing Schools on the Question of Torah and the Sciences in the Rabbinic Academy of the Gra" (Hebrew), *Bekhol Derakheha Da'ehu* 13 (2003): 6–19. But compare Emanuel Etkes, *Ha-Yaḥid be-Doro: Ha-Ga'on mi-Vilnah: Demut ve-Dimui* (Jerusalem: Merkaz Zalman Shazar, 1998), 74–76.

8 My thanks to Dr. Iris Idelson-Shein for her careful demonstration of this copying, and for supplying me with sample passages revealing Hurwitz's indebtedness to Schick: compare Barukh Schick, *Ammudei ha-Shamayim ve-Tifferet Adam* (Berlin, 1776–77), 4b, 21b–22a, 27a, with Pinḥas Hurwitz, *Sefer ha-Brit ha-Shalem*, 242, 244–45.

9 See Iris Idelson-Shein, "'Their Eyes Shall Behold Strange Things': Abraham Ben Elijah of Vilna Encounters the Spirit of Mr. Buffon," *Association for Jewish Studies Review* 36 (2012): 295–322.

10 On Menahem Mendel of Shklov, see Moshe Idel, "The Kabbalah of R. Menahem Mendel of Shklov" (Hebrew), in *Ha-Gra u-Beit Midrasho,* ed. Moshe Ḥallamish, Yosef Rivlin, Raphael Shuchat (Ramat Gan: Bar Ilan University Press, 2003), 176–183; and Yehudah Leibes, "The Students of the Gra, Sabbateanism, and the Jewish Point" (Hebrew), *Da'at,* nos. 50–52 (2003): 255–290.

11 Alan Brill, "Auxiliary to Ḥokhmah: The Writings of the Vilna Gaon and Philosophical Terminology," in Ḥallamish, Rivlin, and Shuchat, eds., *Ha-Gra u-Beit Midrasho*, English section, 9–38, esp. 31–34. My thanks to Rabbi Eliezer Brodt for the reference to Wildman. In his forthcoming book, Eliyahu Stern cites several other members of the Vilna circle who mention Kant and the uncertainty of knowledge, including David Tevle and Abraham Zakheim. They, too, probably consulted Hurwitz's work. My thanks to Professor Stern for sharing with me his unpublished work.

12 Isaac Ḥaver Wildman, *Magen ve-Ẓinnah* (Bnei Brak, Israel, 1985), chap. 6. The reference to *Sefer ha-Brit* is found on 15b, and to Kant on 12b.

13 Brill, "Auxiliary to Ḥokhmah."

14 *Sefer ha-Brit ha-Shalem*, 13.

15 *Sefer ha-Brit* (Brünn, 1797), 4a. I have recently learned that these two individuals are both mentioned, albeit separately, in Jewish communal records from Lemberg as Itsko Minceles and Naḥman Rays. Minceles appears as a communal elder who was sued by the community in 1743. He had borrowed money from two noblemen and from the Jesuits on behalf of the Jewish community but kept it for himself. He was suspended from his public position until he paid back the sum he had taken. Rays, as an agent for a nobleman, was fined along with several other Jews in 1764 for taking funds illegally from the Jewish community. He was a middleman (factor) and the lessee of the service tax called the *krupka,* thus apparently a wealthy man. These incidents are reported by Svjatoslav Pacholkiv in "Gminy żydowskie w Galicji w ll. 1772–1848. Zagadnienia badawcze," in *Galicja 1772–1918: Problemy metodologiczne, stan i potrzeby badań*, ed. Agnieszka Kawalec, Wacław Wierzbieniec, and Leonid Zaszkilniak (Rzeszów, 2011), 2:9–26, esp. 16–17. See also Zbigniew Pazdro, *Organizacya i praktyka żydowskich sądŏw podwojewodzin-*

skich w okresie, 1740–1772 r. (Lwów, 1903), 230–31. My thanks to Professor Rachel Manekin for both references and to Professor Pawel Maciejko for translating the passages from the Polish.

On Jewish Lemberg, see Rachel Manekin, "L'viv," in Gershon Hundert, ed., *The Yivo Encyclopedia of the Jews of Eastern Europe*, ed. Gershon Hundert, www.yivoencyclopedia.org/article.aspx/Lviv. See also Danuta Dombrovska, Abraham Wein, and Aharon Weiss, eds., "Levov/Lwów-Lemberg," in *Pinkas ha-kehilot: Polin,* vol. 2, *Galiẓiyah ha-mizraḥit,* (Jerusalem: Yad va-Shem, 1980), 1–47; Nathan Michael Gelber, ed., *Enẓiklopediyah shel Galuyot,* vol. 4, *Lvov* (Jerusalem and Tel Aviv: Hevrat Enẓiklopediyah shel Galuyot, 1956); and Jerzy Holzer, "'Vom Orient die Fantasie, und in der Brust der Slawen Feuer . . . ': Jüdisches Leben und Akkulturation im Lemberg des 19. und 20. Jahrhunderts," in *Lemberg, Lwów, Lviv: Eine Stadt im Schnittpunkt europäischer Kulturen,* ed. Peter Fässler, Thomas Held, and Dirk Sawitzki (Cologne: Böhlau , 1993), 75–91.

16 See *Sefer ha-Brit ha-Shalem,* 13. On Moses Münz, see Michael Silber, "The Historical Experience of German Jewry and Its Impact on Haskalah and Reform in Hungary," in *Toward Modernity: The European Jewish Model,* ed. Jacob Katz (New Brunswick, N.J.: Transaction Books, 1987), 113.

17 See Rosenblum, "The First Hebrew Encyclopedia," 19–21. Hurwitz pointed out that Lindau's work was copied from Raff in *Sefer ha-Brit ha-Shalem,* 199, more than two hundred years before Tal Kogman allegedly made the same discovery. See Tal Kogman, "Barukh Lindau's 'Resit Limmudim' (1788) and Its German Source: A Case Study of the Interaction between the Haskalah and German 'Philanthropismus,'" *Aleph* 9 (2009): 277–305. Hurwitz explicitly mentions Lindau as well in *Sefer ha-Brit ha-Shalem,* 222 and 227. On Hurwitz's polemic with Maimon, Berlin, and Satanov, see chapter 4 below.

18 This scandal is discussed further in chapter 4.

19 On Ẓevi Hirsch Levin, see the recent summary and bibliography of Hilary Rubenstein, "Lyon, Hart," in *Oxford Dictionary of National Biography,* 2006, www.oxforddnb.com/view/article/17275.

20 *Sefer Brit ha-Shalem,* 1.

21 A fuller discussion of Koerner is found below in chapter 6.

22 The *haskamot* are found in *Sefer ha-Brit ha-Shalem,* 1–7. On the rabbis involved, see my discussion in chapter 1.

23 Rosenblum, "The First Hebrew Encyclopedia," 21–25.

24 *Sefer ha-Brit ha-Shalem,* 570; also cited in Rosenblum, "The First Hebrew Encyclopedia," 22.

25 See *Sefer ha-Brit ha-Shalem,* 163; Rosenblum, "The First Hebrew Encyclopedia," 23–24.

26 I examined the communal ledger located in the municipal archives of The Hague. On this document, see Stefan Litt, *Pinkas, Kahal, and the Mediene: The Records of Dutch Ashkenazi Communities as Historical Sources* (Leiden: Brill, 2008), 11–13.

27 On the Boas family, see I. B. van Crefeld, "De Haagse familie Boas tijdens het ancien régime," *Misjpoge* 10 (1997): 49–66; and Litt, *Pinkas, Kahal, and the Mediene,* 54–58.

28 *Sefer ha-Brit ha-Shalem*, 240.

29 Ibid., 56.

30 The summary of Hart's life and thought that follows is from David Ruderman, "Hart, Eliakim ben Abraham," *Oxford Dictionary of National Biography* (Oxford: Oxford University Press, 2004); idem, "On Defining a Jewish Stance towards Newtonianism: The Case of Eliakim Ben Abraham Hart's *Wars of the Lord*," *Science in Context* 10 (1997): 677–92; and idem, *Jewish Enlightenment in an English Key: Anglo-Jewry's Construction of Modern Jewish Thought* (Princeton: Princeton University Press, 2000).

31 *Sefer ha-Brit ha-Shalem*, 56, 156, 193, and 252, respectively.

32 Rosenblum, "The First Hebrew Encyclopedia," 29.

33 *Sefer ha-Brit* (Brünn, 1797), pt. 2, 66a. See also Rosenblum, "The First Hebrew Encyclopedia," 29–32.

34 On Oppenheim, see Heinrich Flesch, "Oppenheim, Beer ben Isaac (1760–1849)," in *Encyclopedia Judaica*, 2d ed., 15:443; and Michael Brocke, Julius Carlebach, and Carston Wilke, eds., *Biographisches Handbuch der Rabbiner* (Munich: K. G. Saur, 2004), 688–89, no. 1348.

35 *Sefer ha-Brit* (Brünn, 1797), pt. 2, 66a.

36 Rosenblum, "The First Hebrew Encyclopedia," 31–32.

37 Isaac Hirsch Weiss, *Zihronotai: Genazim* (Tel Aviv: Masada Press, 1961), 1:31–32.

38 See my discussion of Hurwitz and Sofer in chapter 6 below.

39 Rosenblum, "The First Hebrew Encyclopedia," 32–33.

40 For a detailed table of contents, see appendix III below.

41 On Fischer, see Iveta Cermanová, "Karl Fischer (1757–1844): The Life and Intellectual World of a Hebrew Censor," *Judaica Bohemiae* 42 (2006): 125–78, and 43 (2007–8): 5–63. The quote is found at 42:177. On Fleckeles and his relationship to Karl Fischer, see also Michael Silber, "Fleckeles, Elazar ben David," in *The Yivo Encyclopedia of Jews in Eastern Europe*, www. yivoencyclopedia.org/article.aspx /Fleckeles_Elazar_ben_David. See also chapter 5 below, n. 19.

42 Professor Michael Silber made the discovery of this letter and I thank him for sharing it with me. Karl Fischer Archives, *Epistolae rabbinorum aliorumque Hebraeorum ab. A. 5549 (1789) usque 5594 (1836) ad me Carolum Fischer etc.*, Národní knihovna ČR (National Library of the Czech Republic), Prague: Call No. XVIII.F.11, fols. 265a–266b. Subsequently, Dr. Iveta Cermanová informed me that she had already examined the letter and added the following details from Fischer's *Tagebuch*: On the ninth of March Fischer received Hurwitz's letter dated March 4, and he answered it on March 12 (bringing the letter to the post office on March 14) and he subsequently filed it (*ad acta*). The full reference is Karl Fischer, *Tagebuch über die Amtsgeschäfte im hebräischen Fache für die Jahre 1788–1805 und 1806–1824*, Národní knihovna ČR [National Library of the Czech Republic], MS, Call No. IX.A.17.a–b, 9, 12, and 14 March 1799. On the last page of the German letter, a Hebrew phrase is added: "*le-yad Morenu ha-rav Eliyahu baal meḥaber Sefer ha-Brit*" (By the hand of our teacher the rabbi Elijah the author of *Sefer ha-Brit*). This would seem to confirm the authenticity of the letter, although it is strange

that Hurwitz's first name "Pinḥas" is left off. I am indebted to Dr. Cermanová for all of these details.

43 "Daher bitte ich demüthigst, mir die Gewogenheit zu erzeigen, Sie möchten die Gnade haben, und Erkundigung bei den dortigen jüdischen Bücher Händlern einzuholen, ob sie wirklich welche hatten, oder noch haben, und von wann sie solche bekommen haben, und die Nachricht schriftlich an die am Ende stehende Adresse zu verabfolgen. Ich will wieder mit Leib und Seele wenn Sie es verlangen gern zu Diensten seyn."

44 *Sefer ha-Brit* (Brünn, 1797), last page, following the list of errors.

45 See *Ha-Me'asef*, 1809, 73 n. 1; see also the fuller discussion of this review in chapter 6 below.

46 *Sefer ha-Brit ha-Shalem*, 17.

47 Ibid., 18–19. Hurwitz explicitly states in the above quote that the publisher of the 1801 pirated edition published only the first part. But there appear to be copies of this edition with two parts, not one. Perhaps there were two separate versions published by Rossmann. For further details, see appendix I on the editions of *Sefer ha-Brit*.

48 The text, which is found in *Sefer ha-Brit ha-Shalem*, 21–23, is translated below in appendix II.

49 *Sefer ha-Brit ha-Shalem*, 21.

50 Ibid., 18.

51 For more on Hurwitz's book as a "commentary" on Vital's manual for attaining prophecy, see chapter 3.

52 I am indebted to Professor Rita Copland for helping me to clarify this point.

53 *Sefer ha-Brit ha-Shalem*, 22.

54 Ibid., 22–23.

55 Ibid., 23.

56 On this edition, the first to appear in an expanded form, see appendix I below.

57 *Sefer ha-Brit ha-Shalem*, 382, where he mentions that he chatted with a Russian army general in Vilna about all matters of warfare being discussed in the Torah; and 384, where he writes from Buczacz that he met an individual named Abba Gabbai who read his work and told him to read Joseph Yabeẓ's *Or ha-Ḥayyim*.

58 See Ḥayyim Friedberg, *Luḥot Zikkaron* (Drohobycz, 1897), 66–67.

59 See, for example, his citations in *Sefer ha-Brit ha-Shalem* of Moses Ḥayyim Luzzatto (435), Joseph Ergas (45, 340, 498, 504), Yosef Delmedigo (47, 299, 314), Sar Shalom Basilea (70–71), Abraham Herrera (141, 143), and Emanuel Ḥai Riki (340, 400).

60 See *Sefer ha-Brit ha-Shalem*, 164, 287, and 300, where he refers to his unpublished works *Sefer Mizvot Tovim*, on the mystical meaning of the commandments; *Sefer Matmonei Mistarim*, on the five secrets in the Book of Daniel; and *Beit ha-Yoẓer*, a commentary on *Sefer Yeẓirah*. His reference to *Ta'am Eẓo* is on 400. He also wrote a short unpublished commentary on one of Abraham Abulafia's works, discussed in chapter 3 below, n. 3.

61 Emanuel Ḥai Riki, *Sefer Mishnat Ḥakhamim*, with the commentary *Ta'am Eẓo* of

Pinḥas Hurwitz (Cracow, 1889). I am referring below to the material at the beginning of the book from page 2 on. Joseph Fischer is briefly mentioned by Kenneth Moss in his essay on printing and publishing after 1800 in the *Yivo Encyclopedia of Jews in Eastern Europe*, www.yivoencyclopedia.org/article.aspx/Printing_and_Publishing/Printing_and_Publishing_after_1800.

62 Mordechai Zalkin, in *Krako-Kaz'imyez'-Krakov: Meḥkarim be-Toldot Yehudei Krakov*, ed. Elhanan Reiner (Tel Aviv: University of Tel Aviv Press, 2001), briefly mentions Solomon. See also Majer Balaban, *Toldot ha-Yehudim be-Krakov uve-Kaz'imyez, 1304–1868* (Jerusalem: Magnes Press, 2002), 2:916.

CHAPTER 3. WHY SHOULD A KABBALIST CARE ABOUT THE NATURAL WORLD?

1 On Hurwitz's kabbalistic sources and other writings, see chapter 2, nn. 59 and 60. On Abulafia, see note 3 below.

2 For a table of contents of the book, see appendix II. Moshe Idel has suggested a typology of three forms of prophecy in Jewish mysticism that are often overlapping: ethical, apocalyptic, and mystical. The mystical focuses on the individual rather than on the community, consisting of the concentration of one's thought and intellect and the divestment from material things. This form of prophecy was particularly prevalent in the eighteenth century prior to and during the emergence of Ḥasidism and was disseminated especially through the many editions of Vital's *Sha'arei Kedushah*. See Moshe Idel, "On Prophecy and Early Hasidism," in *Studies in Modern Religions, Religious Movements, and the Bābī Bahā'ī Faiths*, ed. Moshe Sharon (Leiden: Brill, 2004), 41–75, esp. 47–51.

3 Idel (ibid., 65–67) further claims that Vital's work was directly influenced by the mystical prophecy of Abraham Abulafia, especially in his *Ḥayyei Olam ha-Ba*, which he calls "the most important and detailed handbook for attaining prophecy in eighteenth-century Poland." This observation is particularly relevant in considering Hurwitz's discussion of mystical prophecy, since he too studied Abulafia in his youth, as Idel points out, and even wrote a short commentary on *Ḥayyei Olam ha-Ba*. The composition is available in two manuscripts: Ms. Jerusalem National Library Heb. 8 5403 and Ms. Oxford Bodl. 1584, fols. 1–60, copied in Altona in 1775. See Idel, "On Prophecy and Early Hasidism," 65–67.

4 BT *Avodah Zarah* 20b; with some slight variants, *Mishnah Sotah* 9.15. For additional discussion of *Sha'arei Kedushah* and its notion of attaining prophecy in the present, see R. J. Zwi Werblowsky, *Joseph Karo, Lawyer and Mystic* (Philadelphia: Jewish Publication Society, 1977), 65–77; and Elliot Wolfson, *Through a Speculum That Shines: Vision and Imagination in Medieval Jewish Mysticism* (Princeton: Princeton University Press, 1994), 320–23. See as well the more comprehensive treatment of the book in the unpublished M.A. thesis of Ronit Meroz, "Aspects of Lurianic Discourse of Prophecy" (Hebrew), Hebrew University, Jerusalem, 1980. My thanks to Professor Meroz for sending me a copy of this work.

5 Ḥayyim Vital, *Sefer Sha'arei Kedushah* (Szeged, Hungary, 1903), 1:13a; 2:23b.

6 *Sefer ha-Brit ha-Shalem*, 32–33, 573. See also the thoughtful comments of Resianne Fontaine, "Love of One's Neighbor in Pinḥas Hurwitz's *Sefer ha-Berit*," in *Studies in Hebrew Literature and Jewish Culture, Presented to Albert van der Hade on the Occasion of his Sixty-fifth Birthday*, ed. Martin F. J. Baaesten and Reinier Munk, Amsterdam Studies in Jewish Thought 12 (Dordrecht: Springer, 2007), 294–95, on Hurwitz's fascination with the character of Pinḥas ben Yair.

7 *Sefer ha-Brit ha-Shalem*, 476.

8 Ibid., 499.

9 Ibid., 486. This passage regarding spiritual seclusion and concentrated thought, including the closing of one's eyes, recalls the techniques of *hitbodedut* (self-seclusion) associated with the ecstatic kabbalah of Abraham Abulafia and his sixteenth- century followers. On this, see Moshe Idel, "*Hitbodedut* as Concentration in Ecstatic Kabbalah," in Idel, *Studies in Ecstatic Kabbalah* (Albany: State University of New York Press, 1988), 103–69. The entire text of Hurwitz is translated into English with extensive commentary by David Sears for the website Solitude (*hitboddedut)*, an online archive of translations from classic works and original essays on solitude, meditation, and the inner path in Judaism in memory of Rabbi Aryeh Kaplan (1934–83)": http://solitude-hisbodedus.blogspot.com.es. It is not surprising that the this website singled out this same passage given its interest in spiritual meditation. For a simple commentary on the Lurianic terms "soul-roots," "man of souls," and "action," consult the Sears's commentary on the Solitude website.

10 *Sefer ha-Brit ha-Shalem*, 609.

11 Ibid., 4–5.

12 Ibid., 5.

13 For references to Lindau in *Sefer ha-Brit ha-Shalem*, see, for example, 199, 222, 223; to Tobias Cohen, 54, 89, 91, 92, 183, 290, 484; to Israel Zamosh, 104, 109, 119, 204; to David Gans, 47, 157, 159, 504–5; to Mordechai Shnaber Levinson, 88; to Moshe Ḥefeẓ, 131, 154, 349; to Joseph Delmedigo, 47, 299, 314; and to Abraham Herrera, 141, 143. On passages from Schick that Hurwitz copied, see above, chapter 2, n. 8.

14 Resianne Fontaine, "Natural Science in *Sefer ha-Berit*: Pinchas Hurwitz on Animals and Meteorological Phenomena," in *Sepharad in Ashkenaz: Medieval Knowledge and Eighteenth-Century Enlightened Jewish Discourse*, ed. Resianne Fontaine, Andrea Schatz, and Irene Zweip (Amsterdam: Koninklijke Nederlandse Akademie van Wetenschappen, 2007), 157–81. See also Noah Rosenblum, "Cosmological and Astronomical Discussions in *Sefer ha-Brit*" (Hebrew), *Proceedings of the American Academy for Jewish Research* 62 (1996): 1–36.

15 *Sefer ha-Brit ha-Shalem*, 189.

16 Ibid., 189–90. I discuss Hurwitz's reflections on Kant and his disagreement with Maimon in chapter 4.

17 Ibid., 135.

18 Ibid., 463.

19 Ibid.

20 See ibid., 96–100, 105–8. For a discussion of these subjects in contemporaneous Jewish works, see David Ruderman, *Jewish Thought and Scientific Discovery in Early Modern Europe* (New Haven: Yale University Press, 1995), 332–38; and Fontaine, "Natural Science in *Sefer ha-Berit*," 166–70.

21 *Sefer ha-Brit ha-Shalem*, 115–16.

22 Ibid., 117. Blanchard had moved to London in August 1784 and undertook his first flight over the English Channel to France in January 1785, somewhat later than Hurwitz indicated. For an extensive bibliography of works on and by Blanchard, see http://worldcat.org/identities/lccn-n83-19700.

23 *Sefer ha-Brit ha-Shalem*, 129. On the discovery of the diving bell, see Arthur J. Bachrach, "History of the Diving Bell," *Historical Diving Times* 21 (1998): online at www.thehds.com/publications/bell.html. See also Fontaine, "Natural Science in *Sefer ha-Berit*," 174, who suggests that Hurwitz was referring to Smeaton's invention, which she dates as 1788 and not 1789.

24 *Mishnah Mikva'ot* 10:1 and Bertinora's commentary on this text.

25 *Sefer ha-Brit ha-Shalem*, 129.

26 Another interesting example of Hurwitz's enthusiasm in describing new discoveries is the lightning rod, which he call the *Wetterleiter* (*Sefer ha-Brit ha-Shalem*, 194–96). See Fontaine, "Natural Science in *Sefer ha-Berit*," 170–71; and J. L. Heibron, *Electricity in the Seventeenth and Eighteenth Centures* (Berkeley: University of California Press, 1979), 263. Either Václav Prokop Diviš or Benjamin Franklin was supposedly the inventor of this instrument, but Hurwitz identified Johann Gottlob Krüger of Halle in this role. See also Bernard I. Cohen, "Did Diviš Erect the First European Protective Lightening Rod, and Was His Invention Independent?" *Isis* 43 (1952): 358–64; and Martin Schneider, "Die Elektrizität im Weltbild Johann Gottlob Krügers," *Berichte zur Wissenschaftsgeschichte* 29 (2006): 275–91, who mentions nothing about this alleged invention. On Hurwitz's description of the invention of new acoustical chambers in his day, see *Sefer ha-Brit ha-Shalem*, 278.

27 See chapter 5 regarding his stance on moral cosmopolitanism.

28 On the history of the fight against smallpox, see Genevieve Miller, *The Adoption of Inoculation for Smallpox in England and France* (Philadelphia: University of Pennsylvania Press, 1957); Donald Hopkins, *Princes and Peasants: Smallpox in History* (Chicago: University of Chicago Press, 1983); Derrick Baxby, *Jenner's Smallpox Vaccine: The Riddle of the Vaccinia Virus and Its Origin* (London: Heinemann Educational, 1981); Peter Razzell, *Edward Jenner's Cowpox Vaccine: The History of a Medical Myth* (Firle, Sussex: Caliban Books, 1977); Deborah Brunton, "Pox Britannica: Smallpox Inoculation in Britain, 1721–1830," Ph.D. diss., University of Pennsylvania, 1990; Arnold Rowbotham, "The 'Philosophes' and the Propaganda for Inoculation of Smallpox in Eighteenth-Century France," *University of California Publications in Modern Philology* 18 (1935): 265–90; Andreas-Holger Maehle, "Conflicting Attitudes towards Inoculation in Enlightenment Germany,"*Clio Medica* 29 (1995): 198–222; Adrian Wilson, "The Politics of Medical Improvement in Early Hanoverian London," in *The Medical Enlightenment of the*

Eighteenth Century, ed. Andrew Cunningham and Roger French (Cambridge: Cambridge University Press, 1990), 4–39, and in the same volume, Francis Lobo, "John Haygarth, Smallpox, and Religious Dissent in Eighteenth-Century England," 217–53; Pierre Damon, *La longue traque de la variole: Les pionniers de la médecine préventive* (Paris: Librairie Académique Perrin, 1986); Claudia Huerkamp, "The History of Smallpox Vaccination in Germany: A First Step in the Medicalization of the General Public," *Journal of Contemporary History* 20 (1985): 617–35; and Jonathan B. Tucker, *Scourge: The Once and Future Threat of Smallpox* (New York: Atlantic Monthly Press, 2001).

29 David Ruderman, "Some Jewish Responses to Smallpox Prevention in the Late Eighteenth and Early Nineteenth Centuries: A New Perspective on the Modernization of European Jewry," *Aleph* 2 (2002): 111–44.

30 *Sefer ha-Brit ha-Shalem*, 247.

31 Ibid., 247–48.

32 Ibid., 248–49.

33 Ibid., 249–50.

34 Ibid., 251. Since this section was first included in the 1806–7 edition of Hurwitz's work, the time frame he states would place the discovery of vaccination twenty years earlier than Jenner's publication of 1797, in 1777, which appears inaccurate.

35 Ibid.

36 On Samuelsohn, see Naḥman Gelber, "Toward the History of Jewish Physicians in Poland in the Eighteenth Century" (Hebrew), in *Shai le-Yishayahu: Sefer ha-Yovel le-Yeshayahu Volfsberg*, ed. Israel Tirosh (Tel Aviv: Ha-Merkaz le-Tarbut shel ha-Po'el ha-Mizraḥi, 1956), 361–62; Majer Balaban, *Toledot ha-Yehudim be-Krakov uve-Kaz'imyez, 1304–1868* (Jerusalem: Magnes Press, 2002), 2:749–50, 787, 834–37.

37 Maehle, "Conflicting Attitudes," 219n.20, mentions the propagandistic treatise of Samuel August André David Tissot (1728–97), *L'innoculation justifiée*, published in 1754 in French and in 1756 in German. Hurwitz apparently consulted one of the editions of Tissot's popular medical texts. On these, see Patrick Singy, "The Popularization of Medicine in the Eighteeenth Century: Writing, Reading, and Rewriting Samuel Auguste Tissot's *Avis au peuple sur sa santé*," *Journal of Modern History* 82 (2010): 769–800.

38 *Sefer ha-Brit ha-Shalem*, 251–53.

39 For the fuller story of Herz and the smallpox controversy, see Martin Davies, *Identity or History? Marcus Herz and the End of the Enlightenment* (Detroit: Wayne State University Press, 1995), 117–44.

40 On the controversy over early burial in the German Jewish community, see Moshe Samet, "Burial of the Dead: On the History of the Polemic on Fixing the Time of Death" (Hebrew), *Asufot* 3 (1989–90): 613–65. See also Sigfried Silberstein, "Mendelssohn und Mecklenburg," *Zeitschrift für die Geschichte der Juden in Deutschland* 1, no. 3 (1929): 233–44 and no. 4 (1930): 275–290; Alexander Altmann, *Moses Mendelssohn: A Biographical Study* (Tuscaloosa: University of Alabama Press, 1973), 288–95; Falk Wiesemann, "Jewish Burials in Germany: Between Tradition, the Enlightenment, and the Authorities," *Leo Baeck Year Book* 37 (1992):

17–31; and John Efron, *Medicine and the German Jews: A History* (New Haven: Yale University Press, 2001), 92–104.

41 Ibid., 142–43.

42 Ibid., 144.

CHAPTER 4. JUDAISM AND METAPHYSICS

1 Shmuel Feiner, *The Origins of Jewish Secularization in Eighteenth-Century Europe*, trans. Chaya Naor (Philadelphia: University of Pennsylvania Press, 2010), 244–47, esp. 244–45. In his earlier book, *The Jewish Enlightenment* (Philadelphia: University of Pennsylvania Press, 2002), 348–49, Feiner also referred to Hurwitz as an orthodox Jew. Most recently, one may refer to an essay by Feiner forthcoming in a festschrift in honor of Yosef Kaplan, "*Sefer ha-Brit* Reads the *Haskalah*: A Chapter in the Negation of the Enlightenment at the Turn of the Eighteenth Century" (Hebrew). My thanks to Professor Feiner for allowing me to read this essay prior to publication.

2 For a recent discussion of "orthodoxy" in its eighteenth- and nineteenth-century contexts, with up-to-date bibliography, see David B. Ruderman, *Early Modern Jewry: A New Cultural History* (Princeton: Princeton University Press, 2010), 146–55.

3 *Sefer ha-Brit ha-Shalem*, pt. 1, chap. 20, 318–84.

4 Ibid., 322.

5 Ibid.

6 *Sefer ha-Brit ha-Shalem*, 322–23. I was unable to identify the "milky tube," the sound device Moses allegedly used to broadcast his voice.

7 *Sefer ha-Brit ha-Shalem*, 323.

8 See Jan Assmann, "The Moses Discourse in the Eighteenth Century," chap. 4 in *Moses the Egyptian: The Memory of Egypt in Western Monotheism* (Cambridge, Mass.: Harvard University Press, 1997), 91–143; and Andreas B. Kilcher, "The Moses of Sinai and the Moses of Egypt: Moses as Magician in Jewish Literature and Western Esotericism," *Aries* 4 (2004): 148–70.

9 On the *Traité*, see Sylvia Berti, Françoise Charles-Daubert, and Richard H. Popkin, eds., *Heterodoxy, Spinozism, and Free Thought in Early Eighteenth-Century Europe: Studies on the "Traité des Trois Imposteurs"* (Dordrecht: Kluwer Academic Publishers, 1996).

10 *Sefer ha-Brit ha-Shalem*, 324.

11 Ibid., 330.

12 He undoubtedly relied on Diogenes Laertius or Plutarch or both. On his knowledge of Laertius, see chapter 5.

13 *Sefer ha-Brit ha-Shalem*, 330. He uses several biblical verses in this line. See Nahum 3:15; also see Isaiah 32:2, where he substitutes *kavod*, honor, for *kaved*, heavy, massive.

14 Abraham Melamed, *Rakaḥot ve-Tabaḥot: Ha-Mitus al Mekor ha-Ḥokhmot* (Jerusalem: Haifa University Press, 2010).

15 Satanov is discussed in ibid., 438–40.

16 Ibid., 414–63.

17 On Luzzatto, see also Noah Rosenbloom, *Luzzatto's Ethico-Psychological Interpretation of Judaism* (New York: Yeshivah University Press, 1965). On the parallels between him and Hurwitz with respect to moral cosmopolitanism, see chapter 5 below.

18 *Sefer ha-Brit ha-Shalem*, 335–42.

19 Riki's *Mishnat Ḥakhamim* (Cracow, 1889) was published with Hurwitz's commentary entitled *Ta'am Eẓo* and is discussed in chapter 2.

20 *Sefer ha-Brit ha-Shalem*, 347–57.

21 Ibid., 358.

22 Ibid.

23 Ibid., 360.

24 See Feiner, *The Jewish Enlightenment*, 244–50, 276–83, 335–41; the quote is on 337.

25 Talya Fishman, "Forging Jewish Memory: Besamim Rosh and the Invention of Pre-Emancipation Jewish Culture," in *Elisheva Carlebach*, John M. Efron, and David N. Myers, eds. *Jewish History and Jewish Memory: Essays in Honor of Yosef Ḥayim Yerushalmi* (Hannover and London: University Press of New England, 1998), 70–88.

26 Shmuel Werses, "On Isaac Satanov and his work *Mishlei Asaf*" (Hebrew), in Werses, *Magamot ve-Ẓurot be-Sifrut ha-Haskalah* (Jerusalem: Magnes Press, 1990), 370–92.

27 For more on Berlin and Satanov, see Moshe Pelli, "Isaac Satanov and the Question of Literary Forgery" (Hebrew), *Kiryat Sefer* 54 (1970): 817–24; N. Rezler-Bersohn, "Isaac Satanov: An Epitome of an Era," *Leo Baeck Institute Yearbook* 15 (1980): 81–99; Moshe Samet, "Rabbi Saul Berlin and His Works" (Hebrew), *Kiryat Sefer* 43 (1968): 429–43; and idem, "Rabbi Saul Berlin's *Besamim Rosh*: Bibliography, Historiography, Ideology" (Hebrew), *Kiryat Sefer* 48 (1973): 509–23.

28 Almost all of Hurwitz's comments are based on a few pages of Solomon Maimon's Hebrew commentary on Maimonides, with embellishments as we shall see. His assertion that he cites Kant on the antimonies (see below) from the beginning of *The Critique of Pure Reason* suggests that he never actually saw the book, since the discussion is at the end, not the beginning. My thanks to Professor Yitzhak Melamed for this observation.

29 *Sefer ha-Brit ha-Shalem*, 360.

30 Ibid., 361.

31 Ibid., 362. I have used the modern edition of the *Givat ha-Moreh*, edited by S. H. Bergmann and N. Rosenstreich (Jerusalem: Israel Academy of the Sciences, 1965). For works about Solomon Maimon, see below, note 38.

32 Hurwitz copies from *Givat ha-Moreh*, 16–18.

33 *Sefer ha-Brit ha-Shalem*, 362–64. Hurwitz also mentions Maimon on 189 and 392. For an excellent summary of the philosophy of Leibniz and monadology, with up-to date bibliography, see Brandon Look, "Gottfried Wilhelm Leibniz," *Stanford Encyclopedia of Philosophy*, http://plato.stanford.edu/entries/leibniz/. Kant's

philosophy is masterfully summarized by Paul Guyer in *The Routledge Encyclopedia of Philosophy*, with ample bibliography, http://www.rep.routledge.com/article/DB047SECT1.

34 I cite from the 1929 Norman Kemp Smith translation of *The Critique of Pure Reason*, 29; available at www.hkbu.edu.hk/~ppp/cpr/toc.html.

35 Paul W. Franks, "Jewish Philosophy after Kant: The Legacy of Salomon Maimon," in *The Cambridge Companion to Modern Jewish Philosophy*, ed. Michael L. Morgan and Peter Eli Gordon (Cambridge: Cambridge University Press, 2007), 53–79. See also Jacob Katz, "Kant and Judaism" (Hebrew), *Tarbiẓ* 41 (1971–72): 291–37.

36 David Ellenson, "German Orthodoxy, Jewish Law, and the Uses of Kant," in *The Jewish Legacy and the German Conscience: Essays in Memory of Rabbi Joseph Asher*, ed. Moses Rischin and Raphael Asher (Berkeley: Judah L. Magnes Museum, 1991), 73–84; the quotation is on 80.

37 Alan L. Mittleman, *Between Kant and Kabbalah: An Introduction to Isaac Breuer's Philosophy of Judaism* (Albany: State University of New York Press, 1990), 33.

38 On Solomon Maimon's philosophy, see Abraham P. Socher, *The Radical Enlightenment of Solomon Maimon: Judaism, Heresy, and Philosophy* (Stanford: Stanford University Press, 2006); Samuel Atlas, *From Critical to Speculative Idealism: The Philosophy of Solomon Maimon* (The Hague: Nijhoff 1964); Samuel Hugo Bergman, *The Philosophy of Solomon Maimon* (Jerusalem: Magnes Press, 1967); Meir Buzaglo, *Solomon Maimon: Monism, Skepticism, and Mathematics* (Pittsburgh: University of Pittsburgh Press, 2002); Gideon Freudenthal, *Salomon Maimon: Rational Dogmatist, Empirical Skeptic* (Dordrecht: Kluwer Academic Press, 2003); and Peter Theilke and Yitzhak Melamed "Salomon Maimon (1753–1800)," *Stanford Encyclopedia of Philosophy*, http://plato.stanford.edu/entries/maimon/.

39 See especially *Sefer ha-Brit ha-Shalem*, 392.

40 Ibid., 365.

41 See David Ruderman, *Jewish Thought and Scientific Discovery in Early Modern Europe* (New Haven: Yale University Press, 1995), 353–56 and the bibliography cited.

CHAPTER 5. THE MORAL COSMOPOLITANISM OF PINḤAS HURWITZ

1 See Matthias B. Lehmann, *Ladino Rabbinic Literature and Ottoman Sephardic Culture* (Bloomington: Indiana University Press, 2005), 46–47. Ha-Levi's compilation also includes selections from Solomon Ibn Verga's sixteenth-century work *Shevet Yehudah* and the eighteenth-century *Seder ha-Dorot* of Yeḥiel Heilprin. My thanks to Dr. Katja Smid for this information as well as on the other Ladino editions of Hurwitz's work.

2 Resianne Fontaine, "Love of One's Neighbor in Pinḥas Hurwitz's *Sefer ha-Berit*," in *Studies in Hebrew Literature and Jewish Culture, Presented to Albert van der Hade on the Occasion of his Sixty-fifth Birthday*, ed. Martin F. J. Baaesten and Reinier Munk, Amsterdam Studies in Jewish Thought 12 (Dordrecht: Springer, 2007), 271–95.

3 Hurwitz, *Sefer ha-Brit ha-Shalem*, 565.

4 Ibid., 526.

5 Ibid., 526–27.

6 See esp. ibid., 537.

7 Ibid., 530–31.

8 Ibid., 537–55.

9 Ibid., 556–75.

10 Ibid., 569–70.

11 Ibid., 572–75.

12 Ibid., 230, citing Ḥayyim Vital, *Sefer Sha'arei Kedushah* (Szeged, Hungary, 1903), pt. 1, sh'ar 5, 16a. Compare 32b and 43b, where the verse in Leviticus is evoked. At 25b, however, Vital implies that the verse refers only to Jews. Consider as well Vital's more chauvinistic position regarding the non-Jew in *Likkutei Torah*, as cited and discussed in Elliot R. Wolfson, *Venturing Beyond: Law and Morality in Kabbalistic Mysticism* (Oxford: Oxford University Press, 2006), 112–13. There the non-Jewish nations came into being as a consequence of Adam's transgression. In the state of purity, in other words, Adam is considered a Jew, while in impurity the gentile emerges. Conversion entails liberation of Jewish souls from their entrapment in the body of gentiles.

13 He cites in the same paragraph *Tanna D'vai Eliyahu* 15:26 and the commentary of *Zekukin denurah* (Samuel Heide of Prague, 17th century) on the same text. Both sources acknowledge that the term *neighbor* includes the non-Jew.

14 Ernst Simon, "The Neighbor (*Re'a*) Whom We Shall Love," in *Modern Jewish Ethics*, ed. Marvin Fox (Columbus: Ohio State University Press, 1975), 29–56.

15 Wolfson, *Venturing Beyond*, 4–185.

16 Jacob Katz, *Exclusiveness and Tolerance: Studies in Jewish-Gentile Relations in Medieval and Modern Times* (Oxford: Oxford University Press, 1961), 159–77.

17 See, e.g., Azriel Shochat, *Im Ḥilufei ha-Tekufot* (Jerusalem: Mosad Bialik, 1960), 67–70; David Berger, "Jews, Gentiles, and the Modern Egalitarian Ethos: Some Tentative Thoughts," in *Formulating Responses in an Egalitarian Age*, ed. Marc D. Stern (Lanham, Md.: Rowman & Littlefield, 2005), 83–108; Jay Berkowitz, "Changing Conceptions of Gentiles at the Threshold of Modernity: The Napoleonic Sanhedrin," ibid., 129–50; Edward Breuer, "Jews and Judaism in an Egalitarian Society: Traditionalist Responses in Historical Perspective," ibid., 151–80; Israel Bartal, "Ha-Lo Yehudim ve-Ḥevratam be-Sifrut Ivrit ve-Yiddish be-Mizraḥ Eropah bein ha-Shanim 1856–1914," Ph.D. diss., Hebrew University, 1980; Gerald Blidstein, "Maimonides and Me'iri on the Legitimacy of Non-Judaic Religions," in *Scholars and Scholarship: The Interaction of Judaism with Other Cultures*, ed. Leo Landman (New York: Yeshivah University Press, 1990), 27–35; Moshe Greenberg, "A Problematic Heritage: The Attitude toward the Gentile in the Jewish Tradition—An Israeli Perspective," *Conservative Judaism* 48 (1996): 23–35; David J. Bleich, "*Tikkun Olam:* Jewish Obligations to Non-Jewish Society," in *Tikkun Olam: Social Responsibility in Jewish Thought and Law*, ed. David Shatz, Chaim I. Waxman, and Nathan J. Diament (Northvale, N.J.: Jason Aronson, 1997),

61–102; David Novak, "Gentiles in Rabbinic Thought," in *The Cambridge History of Judaism*, ed. Steven T. Katz, vol. 4 (Cambridge: Cambridge University Press, 2008), 647–62; Laurence J. Silberstein and Robert L. Cohen, eds., *The Other in Jewish Thought and History: Constructions of Jewish Culture and Identity* (New York: New York University Press, 1994); Christine Hayes, "The 'Other' in Rabbinic Literature," in *The Cambridge Companion to the Talmud and Rabbinic Literature,* ed. Charlotte E. Fonrobert and Martin S. Jaffee (Cambridge: Cambridge University Press, 2007), 243–68; and Yitzchak Blau, "The Implications of a Jewish Virtue Ethic," *Torah u-Madda Journal* 9 (2000): 19–41.

18 On his views, see Paweł Maciejko, "Was Rabbi Jonathan Eibeschütz a Crypto-Christian?" paper presented at the Association for Jewish Studies Annual Meeting, December 2007. In it, he cites Karl Anton, Eibeschütz's student who converted to Christianity and wrote a book in German in his defence: "In any case, it must have considerably aggravated the malicious Jews that Rabbi Jonathan did not agree with those [rabbis] who claimed that the doctrine of love of one's neighbour should be restricted only to the Jews, who should care only about other Jews, but claimed that this love is universal in character [*dass diese Liebe allgemein ist*] and that it is as great a sin to cause detriment to people of another religion as to harm members of one's own family."

Maciejko also cites from one of Eibeschütz's sermons in the collection *Ya'arot Devash:* "To be pious is to be purged of all hate; the path to piety is peace. Not to harbor ill feeling to anyone, but to extend kindness even to those who hate us, this is to be virtuous. It is a quality of character demanded by human civilization and dictated in Jewish law; and nothing could so glorify Israel as turning of this trait into a prime instinct of our nature." My thanks to Dr. Maciejko for allowing me to read his paper.

19 Fleckeles is especially important, given his proximity to Hurwitz in both time and place and given his intimate relationship with Karl Fischer, the Prague censor whom Hurwitz addressed in his German letter. Especially relevant is a query raised by Fischer in correspondence with Fleckeles on the binding nature of oaths to non-Jews, which is included in Fleckeles's *Teshuvah me-Ahavah* (Response from Love), 3 vols. (Prague, 1809, 1815, 1821), pt. 1, sec. 26. In his response to his Christian colleague, Fleckeles replied "that the force of an oath is great, and no distinction can be made between making an oath to an Israelite and to a non-Jew." Fleckeles, in fact, was familiar with *Sefer ha-Brit* and cited it in his *Ahavat David* (Prague, 1800), 9a and 25b. I thank Rabbi Eliezer Brodt for the reference. Whether Fleckeles and Hurwitz ever met and shared their universalistic perspectives is a question for future research. For more on Fleckeles and Fischer, see Michael Silber, "Fleckeles, Elazar ben David," in *The Yivo Encyclopedia of Jews in Eastern Europe,* www. yivoencyclopedia.org/article.aspx/Fleckeles_Elazar_ben_David.

20 See Moses Mendelssohn, *Ketavim Ivri'im* 2:3 (*Sefer Netivot ha-Shalom),* in *Gesammelte Schriften*, ed. Werner Weinberg, vol. 17 (Stuttgart: F. Frommann, 1990), 273 (Be'ur commentary on Leviticus 19:18: "This includes all human beings, since all

were created in the image [of God]"). My thanks to Professor Michah Gottlieb for this reference.

21 On Woolf, Michal, and Mecklenburg, see Shochat, *Im Ḥilufei ha-Tekufot;* and Blau, "Implications of a Jewish Virtue Ethic." And on Löwenstamm, see Breuer, "Jews and Judaism in an Egalitarian Society."

22 On Emden, see Shochat, *Im Ḥilufei ha-Tekufot;* Katz, *Exclusiveness and Tolerance;* and Jacob J. Schacter, "Rabbi Jacob Emden, Sabbatianism, and Frankism: Attitudes toward Christianity in the Eighteenth Century," in *New Perspectives on Jewish-Christian Relations, in Honor of David Berger,* ed. Elisheva Carlebach and Jacob J. Schacter (Leiden: Brill, 2012), 359–96. And on Löwenstamm, see Breuer, "Jews and Judaism in an Egalitarian Society."

23 On Aaron of Worms, see especially Berkowitz, "Changing Conceptions of Gentiles at the Threshold of Modernity." More generally, see the new edition of the apologetic treatise of German reform rabbi Elias Grünebaum, originally published in 1867, entitled *Die Sittenlehre des Judentums,* ed. Carsten Wilke (Vienna: Bölow, 2010). In this work (281–85), Grünebaum provided an anthology of early modern Jewish authors writing on the subject of loving one's neighbor, based on Eliezer Landshuth's preface to the *Siddur Hegjon ha-Lev* (Koenigsberg, 1845). Finally, see the recent essay of Asaf Yedidya, "Between Reform and Apologetics: The Public Letter of Isaac Markus Jost on 'Love Thy Neighbor' and Rabbi Zvi Hirsch Kalischer's Response," *Zutot* 10 (2012): 1–11.

24 Indeed, Luzzatto and Benamozegh follow an even longer tradition of tolerance toward the non-Jew among Italian Jewish thinkers. These include Jacob Saraval, *Lettera Apologetica* (Mantua, 1775); and Elia Morpurgo, *Discorso* (Gorizia, 1783). My thanks to Dr. Asher Salah for these references. And even earlier, one might point to expressions of tolerance in the writings of Ovadia Sforno, Leon Modena, Mordechai Dato, and Judah del Bene. On these, see Talya Fishman, "Changing Early Modern Discourse about Christianity: The Efforts of Rabbi Leon Modena," in *Ari Yishag: R. Aryeh Yehudah Modena ve-Olamo,* ed. David Malkiel (Jerusalem: Makhon Ben Ẓevi, 2003), 184–86.

25 See Marc Gopin, "An Orthodox Embrace of Gentiles? Interfaith Tolerance in the Thought of S. D. Luzzatto and E. Benamozegh," *Modern Judaism* 18 (1998): 173–95. The quotation is from 176, translated from Luzzatto, *Il Giudaismo Illustrato* (Padua, 1848), 11. See also Marc Gopin, "The Religious Ethics of Samuel David Luzzatto," Ph.D. diss., Brandeis University, 1992, esp. 339–48.

26 Quoted in Gopin, "An Orthodox Embrace," 180; translated from Luzzatto's *Lezioni di teologia morale israelitica* (Padua, 1862), 33–34.

27 Gopin, "An Orthodox Embrace," 182–92. Hurwitz's moral stance might also be compared favorably with that of Abraham Isaac Kook. On Kook's directive to love all humanity, see Blau, "Implications of a Jewish Virtue Ethic," 35–36; and see Benjamin Ish Shalom and Shalom Rosenberg, eds. *The World of Rav Kook's Thought* (Jerusalem: Avi Chai Foundation, 1991), 207–54, 423–35.

28 Hurwitz cites both Riki and Ergas extensively in *Sefer ha-Brit ha-Shalem*, e.g. 45,

340, 400, 498, and 504. On Hurwitz's commentary on Ḥai Riki's work, see chapter 2 above, and see my discussion of Hurwitz as expositor of kabbalah in chapter 3.

29 See, for example, Abraham Melamed, "The Hebrew Encyclopedias of the Renaissance," in *The Medieval Hebrew Encyclopedias of Science and Philosophy*, ed. Steven Harvey (Dordrecht: Kluwer Academic Publishers, 2000), 441–64; and more generally, David Ruderman, "At the Intersection of Cultures: The Historical Legacy of Italian Jewry Prior to the Emancipation," in *Gardens and Ghettos: Art and Jewish Life in Italy*, ed. Vivian Mann (Los Angeles: University of California Press, 1989), 1–23.

30 *Sefer ha-Brit ha-Shalem*, 529–30.

31 Ibid.

32 On Leopold's life, see Paul Zimmermann, "Leopold (Maximilian Julius Leopold), Herzog zu Braunschweig," *Allgemeine Deutsche Biographie* 18 (Leipzig, 1883): 376ff.

33 Leopold's death and its aftermath are the subject of a recent book by Anton Pumpe, *Heldenhafter Opfertod des Herzogs Leopold von Braunschweig 1785 in der Oder: Wahrheit oder Legende? Presse in Spannungsfeld zwischen Aufklärung und Propaganda: Eine quellenkritische Studie* (Braunschweig: Braunschweigischen Landesmuseums, 2008). It contains a very extensive bibliography. Also useful is Gert-Dieter Ulferts, "Denkmale für einen Helden der Aufklärung: Bildkünstlerische Reaktionen auf den Tod Herzog Leopolds von Braunschweig 1785," in *Braunschweig-Bevern: Ein Fürstenhaus als europäische Dynastie, 1667–1884*, ed. Christof Römer (Braunschweig: Braunschweigischen Landesmuseums, 2007), 465–78.

34 Quoted in translation by Lady Blennerhassett, *Madame de Staël: Her Friends and Her Influence in Politics and Literature* (London, 1889), 2:83.

35 Ibid., 83–84.

36 See Ulferts, "Denkmale für einen Helden"; and Pumpe, *Heldenhafter Opfertod*.

37 This translation of the original 1785 edition may be found at http://raptusassociation.org/ode1785.html.

38 Robert Beachy, "Recasting Cosmopolitanism: German Freemasonry and Regional Identity in the Early Nineteenth Century," *Eighteenth-Century Studies* 3 (2000): 266–67.

39 The literature on Lessing is of course daunting. A useful summary is Henry E. Allison, *Lessing and the Enlightenment: His Philosophy of Religion and Its Relation to Eighteenth-Century Thought* (Ann Arbor: University of Michigan Press, 1966). See also H. B. Nisbet, ed., *Lessing: Philosophical and Theological Writings* (Cambridge: Cambridge University Press, 2005).

40 On Mendelssohn and Lessing, see Alexander Altmann, *Moses Mendelssohn: A Biographical Study* (Tuscaloosa: University of Alabama Press, 1973).

41 On Jacob Hart, see chapter 2.

42 For example, *Sefer ha-Brit ha-Shalem*, 339, on the superiority of Jewish faith; or 388–89, on the kabbalists' distinction between Jews and non-Jews.

43 On this cosmopolitanism, see, e.g., Pauline Kleingeld, "Six Varieties of Cosmopolitanism in Late Eighteenth-Century Germany," *Journal of the History of Ideas* 60 (1999): 505–24; Thomas J. Schlereth, *The Cosmopolitan Ideal in Enlightenment Thought: Its Form and Function in the Ideas of Franklin, Hume, and Voltaire, 1694–1790* (Notre Dame: Notre Dame University Press, 1977); James Bohman and Matthias Lutz-Bachmann, eds., *Essays on Kant's Cosmopolitan Ideal* (Cambridge, Mass.: Harvard University Press, 1997); and the essay on cosmopolitanism by Pauline Kleingeld and Eric Brown in the *Stanford Encyclopedia of Philosophy*, http://plato.stanford.edu/entries/cosmopolitanism.

44 On Teomim, see Alfassi, Itzhak, "Teomim, Joseph ben Meir," *Encyclopedia Judaica*, 2d ed., vol. 19 (Detroit: Macmillan and Keter Publishing House, 2009), 643–44 and the older bibliography he cites. The sermon of Teomim was printed by Johann Christian Winter in 1785. I found a copy in the University of Chicago Library.

45 Joseph bem Meir Teomim, *Leichenrede des Herzogs Maximilian Julius Leopold von Braunschweig* (Frankfurt an der Oder, 1785), 6, 8, 14, 12, 12, 13, respectively.

46 Ibid., 5, 9, 11, respectively.

47 Ibid., 14–16.

48 Naphtali Herz Wessely, "Duke Leopold: A Living and Most Accomplished Person" (Hebrew), *Ha-Me'asef* 2 (1785): 145–52; the quotation is on 152. My thanks to Professor Shmuel Feiner for calling my attention to this source.

49 For an additional contact that Leopold had with a Jew, see Lea Ritter-Santini, "Die Erfahrung der Toleranz: Melchisedech in Livorno," *Germanisch-Romanische Monatsschrift* 47 (1997): 317–62. The article describes the brief visit of Leopold, accompanied by his teacher, Lessing, to Livorno between March 13 and 18, 1775. According to a later biographer, the two engaged in conversation with the local rabbi, probably Abraham Isaac Castello, at the synagogue. Lessing referred to the latter as being even more profound a metaphysician than Mendelssohn. The author claims that this meaningful set of conversations, in the special ambiance of the Italian Renaissance and especially Boccaccio and the *Decameron*, might have influenced Lessing in his views on toleration between Jews, Moslems, and Christians. My thanks to Professors Francesca Bregoli and Asher Salah for this reference. Cf. also Meyer Kayserling, "Herzog Leopold von Braunschweig und die Juden" *Jeschurun* 4 (1858): 308–14.

50 *Sefer ha-Brit ha-Shalem*, 540, 541, 555, and 563.

51 See Fontaine, "Love of One's Neighbor," 291–93.

52 Hurwitz's citation on 563 seems to refer to the life of Bias in Diogenes Laertius, *Lives and Opinons of Eminent Philosophers*, transl. C. D. Yonge, book 1, http://classicpersuasion.org/pw/diogenes/dlbias.htm: "It was a saying of his that is was more agreeable to decide between enemies than between friends; for that of friends, one was sure to become a friend."

53 *Sefer ha-Brit ha-Shalem*, 446. The story appears in various works, including Cicero, which does not seem to be Hurwitz's source. Schiller's *Die Bürgschaft* was based on a version in the *Gesta Romanorum*.

54 *Sefer ha-Brit ha-Shalem*, 570–72. I have followed the JPS English translation of the *Tanakh* (Philadelphia, 1985).

55 See B. T. *Sukkah* 5b; *Ḥagigah* 13b.

56 Rashi on Exodus 26:31.

57 Isaac Abravanel on Exodus 25:10 (*Perush al ha-Torah* [Jerusalem, 1964], 152).

58 Bracha Yaniv, "The Cherubim on Torah Art Valences," *Assaph*, sec. B4 (1999): 155–70.

59 *Sefer ha-Brit ha-Shalem*, 571.

60 Ibid.

61 See, e.g., George Oliver's chapter on the cherubim in his *Signs and Symbols Illustrated and Explained in a Course of Twelve Lectures on Freemasonry* (London, 1838), 61–80. See also David B. Ruderman, "The Mental Image of Two Cherubim in Pinḥas Hurwitz's *Sefer ha-Brit*: Some Conjectures," in *Festschrift in Honor of Richard Cohen*, forthcoming, where I offer more iconographic support for this suggestion.

62 Resianne Fontaine reminds me that in her article on Hurwitz's scientific sources ("Natural Science in *Sefer ha-Berit:* Pinchas Hurwitz on Animals and Meteorological Phenomena," in *Sepharad in Ashkenaz: Medieval Knowledge and Eighteenth-Century Enlightened Jewish Discourse*, ed. Resianne Fontaine, Andrea Schatz, and Irene Zweip [Amsterdam: Koninklijke Nederlandse Akademie van Wetenschappen, 2007], 157–81), she noted that Hurwitz emphasized the social life and cooperation of beavers and suggests that this reflection might be meaningfully connected to his passionate call for loving all persons. I think this is an important observation linking his science to his moral concerns. Appreciating the natural world around us certainly might make us more aware of our collective humanity and our social responsibility to all creatures, both humans and animals.

CHAPTER 6. THE READERS OF *SEFER HA-BRIT*

1 See appendix 1 for a detailed inventory of the editions, translations, and abridgments of *Sefer ha-Brit*.

2 Part 1 was published in Berlin, 1788; Brünn, 1796; and Cracow, 1820. Part 2 appeared in Dessau in 1810.

3 *Ha-Me'asef*, 1809, 68–75, 136–39. The review is well summarized in Ben Zion Katz, *Rabbanut, Ḥasidut, Haskalah: Le-Toledot ha-Tarbut ha-Yisraelit me-Sof ha-Me'ah Ha-16 ad Reishit ha-Ma'ah ha-19* (Tel Aviv: Devir Publishing House, 1956–58), 2:148–51. The journal stopped publication in Breslau in 1797, with only four issues of volume 7 appearing, and only resumed in 1809 in Berlin, Altona, and Dessau. It is evident that the reviewer saw only the first edition of Hurwitz's book, most likely soon after it appeared in 1797, though could not publish the review until 1809.

4 This is the suggestion of Noah Rosenblum, "The First Hebrew Encyclopedia: Its Author and Development" (Hebrew), *Proceedings of the American Academy for Jewish Research* 55 (1988): 59n.1.

5 *Ha-Me'asef*, 1809, 69.

6 Ibid. If Wolfsohn-Halle was indeed the reviewer, this line might offer direct evidence that Hurwitz visited Breslau, where Wolfsohn-Halle taught in the Jewish school for a long period of his life.

7 Ibid., 73n.1.

8 Ibid., 70.

9 Ibid., 74–75.

10 Ibid., 136–39.

11 On Koerner and his other publications, see *Encyclopaedia Judaica*, 2nd ed. (Detroit: Macmillan, 2007), 12:254.

12 Moses Koerner, *Ke'Or Nogah* (Breslau, 1816), title page and 1. On Hurwitz's connections with Rabbis Ẓevi Hirsch Levin and Saul Loewenstamm, see chapter 2 above.

13 *Ke'Or Nogah*, 1.

14 Ibid., 34.

15 Ibid., 46.

16 David Caro and Judah Leib Miesis, *Sefer Tekhunat ha-Rabbanim* (Vienna, 1823), 25, second note. On Caro, see *Encyclopaedia Judaica* 5:192–93; on Mieses, 11:1527. See also Shmuel Feiner, *Milḥemet Tarbut: Tenu'at ha-Haskalah ha-Yehudit be-Ma'ah ha-19* (Jerusalem: Merkaz Zalman Shazar, 2010), 122–25 and index; Robert Katz, "David Karo's Analysis of the Rabbi's Role," *Central Conference of American Rabbis Journal* 13 (1966): 41–46; and Avner Holtzman, www.yivoencyclopedia.org/article.aspx/Mieses_Yehudah_Leib.

17 On Nathan Sternharz of Nemirov, see Feiner, *Milḥemet Tarbut*, 122–25.

18 *Te'udah Be-Yisra'el* (Vilna, 1828), end of chap. 20, 100. On Levinsohn, see Mordechai Zalkin, www.yivoencyclopedia.org/article.aspx/Levinzon_Yitshak_Ber.

19 *Bikkurei Ribal* (Warsaw, 1891), 28–29.

20 Samuel Joseph Fuenn, *Kiryah Ne'emanah* (Vilna, 1860), 202–4. On Fuenn, see *Encyclopaedia Judaica* 7:213–14.

21 Fuenn, *Kiryah Ne'emanah*, 204. See also my remarks on Fuenn's account in chapter 2.

22 Eliezer Zweifel, *Sanegor* (Warsaw, 1885), 268, 276. On Zwiefel, see Feiner, *Milḥemet Tarbut*, 150–180; Shmuel Feiner, *Haskala ve-Historia* (Jerusalem: Merkaz Zalman Shazar, 1995), 416–30; Gloria W. Pollack, "Eliezer Zvi HaCohen Zweifel: Forgotten Father of Modern Scholarship on Hasidism," *Proceedings of the American Academy for Jewish Research* 49 (1982): 87–115; and *Encyclopaedia Judaica* 16:1245–46. On Alkalai, see *Encyclopaedia Judaica* 2:637. Zweifel's source is Abraham Alkalai, *Sefer Zakhor le-Avraham* (Salonika, 1798; 2d ed. Munkatch, 1895), pt. 3, letter 9.

23 This of course is only a small sampling of Maskilim who were familiar with Hurwitz's work. Mordechai Aaron Guenzburg (1795–1846), a Hebrew-language writer and the founder of the first modern Jewish school in Lithuania, relates in his autobiographical work *Avi'ezer* that he had read *Sefer ha-Brit* and the *Phaedon* of Moses Mendelssohn by the time he reached the age of seventeen. See Israel Bartal, "Mordechai Aaron Guensburg: A Lithuanian Maskil in the Face of Moder-

nity," in Emanuel Etkes, *Ha-Dat ve-ha-Ḥayyim: Tenua't ha-Haskalah ha-Yehudit be-Mizraḥ Eropah* (Jerusalem: Merkaz Zalman Shazar, 1993), 111.

We might also add, in the mold of Fuenn and Zweifel, the Romanian rabbi and biblical commentator Meir Leibush Michel Weiser (1809–79), known as the Malbim. Noah Rosenblum, in his *Ha-Malbim: Parshanut, Pilosofia, Madah u-Mistorin be-Kitvei ha-Rav Meir Levush Malbim* (Jerusalem: Mosad ha-Rav Kuk, 1988), 207–77, offers a full discussion of the Malbim's reliance on *Sefer ha-Brit*, including his reference to Kant, which seems to be taken from Hurwitz. Rosenblum also suggests (259) that Hurwitz influenced another like-minded enlightened traditionalist, Jacob Ẓevi Mecklenburg (1831–65), in his most important work *Ha-Ketav ve-ha-Kabbalah* (Leipzig, 1839). Samuel Alexandrov, discussed below, in his *Masekhet Nega'im* (Warsaw, 1886), 25, had already suggested that Mecklenburg copied from *Sefer ha-Brit* without attribution.

Raphael Ze'ev ha-Kohen, the author of *Ḥut ha-Meshullash* (Odessa, 1874), strongly criticized the notion that Israel was the source of all knowledge, based on his reading of *Sefer ha-Brit*. He is discussed in Shmuel Werses, *Haskalah ve-Shabta'ut* (Jerusalem: Merkaz Zalman Shazar, 1988), 142–43. The rabbinic scholar and Maskil Shlomo Yehudah Rappaport (1790–1867), known by the acronym Shir, was also familiar with Hurwitz's book. See Isaac Barzilay, *Shlomo Yehudah Rapoport (Shir), 1790–1867, and His Contemporaries* (Ramat-Gan: Masada Press, 1969), 29.

24 In writing this section, I am greatly indebted to Dr. Maoz Kahana, who pointed me to all the sources on Ḥatam Sofer and Hurwitz discussed here. I offer my sincere thanks for his assistance.

25 Solomon Sofer, *Ḥut ha-Meshullash* ([1887] Drohobicz, 1908), 29a. This comment was added by the author to the second expanded edition (Mukachevo, 1893). On Ḥatam Sofer, see the Hebrew University dissertation of Maoz Kahana, "Bein Prague le-Pressburg: Ketivah Hilkhatit be-Olam Mishtaneh me *ha-Nodah be-Yehudah* ad *ha-Ḥatam Sofer*, 1730–1839," soon to be published in book form.

26 Moses Sofer, *Kovez̧ Teshuvot Ḥatam Sofer* (Jerusalem: Makhon Ḥatam Sofer, 1999), no. 26. First published by Eliezer Zusman Sofer in *Et Sofer* (Pakś, 1887–88), it was later placed in the full collection of Sofer's responsa. An earlier version on astronomical matters appeared in *Ḥiddushei Ḥatam Sofer al Shulkhan Arukh Yoreh De'ah* (London, 1955), siman 116, 18.

27 On Hurwitz's sojourn in Pressburg and his relationship with Oppenheim, see chapter 2.

28 Hurwitz, *Sefer ha-Brit* (Brünn, 1797), pt. 1, ma'amar 4, end of chap. 10, 22a.

29 Sofer, *Ḥiddushei Ḥatam Sofer al Massekhet Tamid* (Jerusalem?, 1942), 179.

30 Hurwitz, *Sefer ha-Brit* (Vilna, 1817), pt. 1, ma'amar 4, end of chap. 10, 72a–73a.

31 Sofer, *Ḥiddushei Ḥatam Sofer al Massekhet Tamid*, 179.

32 Upon reading this section, Dr. Kahana offered me the following thoughtful comments, with which I fully concur: "I think you should make clear that this exchange was not mediated by print, and reflects a live direct connection. You should also remember that Ḥatam Sofer was neither 'orthodox' nor famous in 1799. At this time he was thirty-seven years old, Rabbi of Mattersburg (where he

arrived just one year earlier), clearly unknown out of his Burgerland region. . . . Therefore the analysis of power relations and 'adaption' to rabbinical authority seems to me a little bit anachronistic. The questions here are the general motivation [of Hurwitz in making] corrections and expansions in the second and third editions, as well as . . . [his] uses of the [notion of] 'secret (*sod*)' as a solution to the validity of outdated traditional scientific knowledge, while managing the complicated reciprocities between science and kabbalah. This last point seems to me crucial."

Dr. Kahana recently completed another essay related to this subject entitled "The Exorcist from Prague: A Chapter in the Scientific Thinking of Moses Sofer" (Hebrew), to appear in a future issue of the *Association for Jewish Studies Review*. In this study, Kahana refers to Sofer's astronomical treatise, which he left unfinished after reading and appreciating *Sefer ha-Brit*, and his general reliance on Hurwitz's book when dealing with natural history. More importantly, in his discussion of Sofer's treatment of demonology, Kahana shows the parallels between Sofer and Hurwitz with respect to their integration of kabbalah and science, their appreciation of the power of magic, and their eclectic approaches to treating the allegedly mysterious forces of the natural world. Kahana points out Hurwitz's inconsistencies in his perception of demons, at once acknowledging their existence and dismissing them, but he shows how the two men shared, along with other Jewish thinkers, a common mental universe with respect to magic and science at the beginning of the nineteenth century.

33 Akiba Eger, *Haga'ot Rabeinu Akiva Eger* (Berlin, 1862), Hilkhot Rosh Ḥodesh, siman 426, se'if alef.

34 Akiba Schlesinger, *Sefer Ẓeva'at Moshe* (Vienna, 1863), na'ar ivri, 16a–b.

35 *Sefer ha-Brit*, pt. 1, ma'amar 6, chap. 3, 89.

36 The following discussion revolves around Abraham ben Yeḥiel Danzig, *Binat Adam*, sha'ar issur ve-heter, siman 34, 49–50 (from Bar Ilan University Responsa online). On Danzig, see *Encyclopaedia Judaica* 5:1297–98. On this responsum, see J. David Bleich, *Bioethical Dilemmas: A Jewish Perspective* (Hoboken, NN.J.: Ktav Publishing House, 1998), 236–37.

37 Ẓevi Hirsch Shapira, *Beit Yisra'el*, Yoreh De'ah, pt. 2 (Vilna, 1892), siman 7. Ẓevi Hirsch was the father of Ḥayyim Elazar Shapira. On the latter, see Alan Nadler, "The War on Modernity of R. Ḥayyim Elazar Shapira of Munkacz," *Modern Judaism* 14 (1994): 233–64.

38 Israel Abraham Alter Landau, *Beit Yisra'el*, Yoreh De'ah (Brooklyn, 1994), siman 7, 15–17.

39 J. H. Beck-Cohen, *Sefer Beit Shmuel*, pt. 2 (Jerusalem, 1961), daltei teshuvah, 408.

40 Elijah be Yosef Avirazal, *Dibrot Eliyahu*, pt. 4 (Jerusalem 1993), siman 9, 30–31.

41 Ovadiah Yosef, *Yehaveh Da'at*, pt. 6, siman 47 (from Bar Ilan University Responsa online). This opinion is discussed by Daniel Sperber in www.biu.ac.il/JH/Parasha/eng/sukot/spe.html.

42 Menashe Klein, *Mishneh Halakhot*, pt. 4, siman 129 (from Bar Ilan University Responsa online). Beyond the specific issue of worms in vinegar, Hurwitz's book,

of course, was consulted on a variety of issues by other orthodox rabbis. See, e.g., Yaakov Yisrael Kanievsky (1899–1985), the so-called Steipler Gaon, in his *Sefer Ḥayyei Olam* (Bnai Brak, 1957), 33–37, who cites *Sefer ha-Brit* extensively on the instability of the sciences, the wonders of nature, Hurwitz's critique of philosophy, and even Kant. Shalom Mordechai ha-Cohen Schwadron (1835–1911), the well-know Galician halakhic authority, in his *Shut Maharsham*, pt. 7, siman 103 (from Bar Ilan University Responsa online), quotes Hurwitz on the conditions of boat travel in modern times when discussing the status of a body lost at sea. Israel Shlomo Zalman Alexandrovsky (1810–77), the rabbi of Lyakhavichy, Belarus, was impressed by Hurwitz's discussion of kabbalistic study; see his *Sefer Iggeret ha-Ḥayyim* (Warsaw, 1911), 13, 31. Rabbi Ẓevi Hirsch Kalisher, in his *Emunah Yeshara* (Krotosehin, 1843), pt. 1, 40a, cites Hurwitz on the relation of faith and reason. My thanks to Dr. Eliezer Brodt for this last reference.

Other rabbis who consulted Hurwitz include Samuel (1794–1872) and Mattityahu Strashun (1817–85), the famous Vilna rabbi and his learned maskilic son. See Shua Engelman, "Ha-Rav Shmuel Strashun (Harashash) ve-Hagahotov le-Talmud Bavli," Ph.D. diss., Bar Ilan University, 2009, 236; Mattityahu Strashun, *Mivḥar Ketavim* (Jerusalem: Mosad ha-Rav Kuk, 1969), 140; and the essay by Mordechai Zalkin on both at www.yivoencyclopedia.org/article.aspx/Strashun_Shemuel_and_Matityahu. The Galician rabbi Meshulam Roth (1875–1962) even included *Sefer ha-Brit* in his ambitious curriculum of Jewish studies; see his *Mevasser Ezra* (Brooklyn, 1993), 177. Rabbi Meir Yonah Shatz (d. 1891), in his *Mei ha-Shilo'ah* (Brooklyn, 1883), 45a–b, also cited Hurwitz's work. My thanks to Rabbi Eliezer Brodt for these last three references. There are undoubtedly many more.

43 Despite his attack on the Sabbateans, Wolf, the son of Jonathan Eibeshütz, appears to have been familiar with Hurwitz's book and cited it; see *Yehudah Liebes* (Jerusalem: Mosad Bialik, 1995), 349n.168.

44 *Sefer ha-Brit ha-Shalem*, 375–76.

45 Moshe Idel, *Kabbalah: New Perpectives* (New Haven: Yale University Press, 1988), 152–53 (including the quote of Hurwitz).

46 Mendel Piekartz, *Ḥasidut Bratslav: Perakim be-Ḥayyei Meḥolela u-ve-Ketaveha* (Jerusalem: Mosad Bialik, 1995), 249–52.

47 Zvi Mark, *Mistika u-Shiga'on be-Yeẓirat R. Naḥman mi-Breẓlav* (Tel Aviv: Am Oved, 2003), 87-88, esp. nn. 9 and 10.

48 Joseph Perl, *Megalleh Temirim* (Vienna, 1891), letter 104. I am indebted to Professor Jonatan Meir for this source.

49 Isaac Satanov, *Ḳunṭres mi-Sefer ha-Zohar, Ḥibura Tinyana* (Berlin, 1783), 12b–13a. My thanks to Professor Jonatan Meir for this reference.

50 Heikel Horowitz, "The More Knowledge, the More Peace" (Hebrew), *Ha-Boker Or* 1 (1876): 361–64. My thanks to Professor Jonatan Meir for this reference. The letter is also cited by Shmuel Feiner in *The Origins of Jewish Secularization*, trans. Chaya Naor (Philadelphia: University of Pennsylvania Press, 2010), 110.

Maoz Kahana kindly supplied me with the following Ḥasidic source: "On one occasion, it was told to the Rav [Rabbi Pinḥas of Koretz (1728–1790)] in the name of one book whose name I don't presently know that one needs to love a gentile because he is a divine creation, and the rabbi appreciated this very much" (*Imrei Pinḥas* [Jerusalem, 2001], 415). The unidentified book is undoubtedly Hurwitz's work, and although the rabbi himself would not have been alive after its publication, one of his disciples probably referred to Hurwitz's chapter. The source is important in illustrating how so radical a moral stance could be embraced within traditional Ḥasidic circles.

One more indication of the popularity of *Sefer ha-Brit* in Ḥasidic circles is the publication of the Yiddish edition in 1928–29 in Warsaw by Avraham Yosef Klaiman. In a personal communication, Professor Yitzhak Melamed told me that he saw this edition included as part of a larger work called *Maẓmiaḥ Yeshuot,* comprising four parts and compiled by a certain Mendel Ravitski. The volume includes a long list of approbations by Ḥasidic rabbis (Sadigura, Viznitz, Munkatch, and others). This was clearly a Ḥasidic book, and it appears that Ravitski considered *Sefer ha-Brit* worthy of inclusion in such a volume.

51 *Sefer ha-Brit ha-Shalem*, 17.

52 See Appendix I for these editions and other bibliographical references.

53 Eliezer Papo, *Pele Yo'eẓ* (Constantinople, 1824), 3. On his use of Hurwitz in the Ladino version, see the references collected in Matthias B. Lehmann, *Ladino Rabbinic Literature and Ottoman Sephardic Culture* (Bloomington: Indiana University Press, 2005), 195, 198–200. See also Matthias Lehmann, "Representation and Transformation of Knowledge in Judeo-Spanish Ethical Literature: The Case of Eli'ezer and Judah Papo's *Pele Yo'ets*," in *Jewish Studies between the Disciplines*, ed. Klaus Hermann et al. (Leiden: Brill, 2003), 299–324; and Marc Angel, ed., *Eliezer Papo: The Essential Pele Yoetz* (Brooklyn, N.Y.: Sepher-Hermon Press, 1991).

54 Raphael Kazin, *Sefer Derekh ha-Ḥayyim* (Constantinople, 1848), Sha'ar ha-Emunah, chaps. 8–12, 30a–35a, and chap. 19, 40a–41a; Sha'ar ha-Tokhahot, chap. 10, 48a–b. For the context of this work, see Yaron Harel, "Likkutei Amrim in Ladino: On the Polemical Literature of Rabbi Raphael Kazin" (Hebrew), in *Languages and Literatures of Sephardic and Oriental Jews, Proceedings* 6 (2009): 106–19.

55 Moses Pardo, "Response Out of Love" (Hebrew), *Ha-Levanon*, October 16, 1872, 10. On Moses Pardo, see *Encyclopaedia Judaica* 13:93–94.

56 Ben Ish Ḥai, *Shut Rav Pa'alim*, p.t 2 Orakh Ḥayyim, siman 1 (from Bar Ilan University Responsa online); *Sefer Ben Yehoyada* (Jerusalem, 2007), Berakhot, chap. 6, 42a; Pesaḥim, end of chap. 9. On Ben Ish Ḥai, see *Encyclopaedia Judaica* 10:242–43.

57 Ben Ish Ḥai, *Shut Rav Pa'alim*, pt. 2 Sod Yesharim, siman 3 (from Bar Ilan University Responsa online).

58 Isaac Akrish, *Sefer Kiryat Arbah* (Jerusalem, 1876), 186a. On Akrish, see Joseph Ringel, "Arkish, Isaac ben Abraham," *Encyclopedia of Jews in the Islamic World*, ed. Norman Stillman (Leiden: Brill, 2012), http://referenceworks.brillonline.com/entries/encyclopedia-of-jews-in-the-islamic-world/akrish-isaac-ben-abraham-SIM_000300.

59 Shalom Hedaya, *Maḥberet Shalom le-Am* (Jerusalem, 1929), pt. 2, chap. 6, 35a. On Shalom and the circle of Beit El, see Pinḥas Giller, *Shalom Shar'abi and the Kabbalists of Beit El* (Oxford: Oxford University Press, 2008), esp. 86.

60 Raḥamim David Sarim, *Sefer Sha'arei Raḥamim* (Jerusalem, 1926), 3. On Sarim, see Giller, *Shalom Shar'abi*, 87–88.

61 Ḥayyim Blia'h, *Sha'ar Kevod Adonai* (Jerusalem, 1986), 10.

62 Maẓli'aḥ Mazouz, *Sefer Ish Maẓli'aḥ* (Jerusalem, 1973), Orekh Ḥayyim, siman 13, 51a.

63 Pinḥas Zabiḥi, *Sefer Tehillat Pinḥas* (Jerusalem, 2010), 101–3.

64 See Rachel Morpurgo, *Ugav Raḥel* (Cracow, 1903), 107. My thanks to Professor Tovah Cohen for this reference. Cohen has just completed a monograph on Morpurgo entitled *Ugav Ne'elam: Masah be-Ikvot ha-Meshoreret ha-Ivriyah Italkiyah Raḥel Morpurgo,* where she discusses Hurwitz's impact on her writing. See also Marina Arbib, "Una voce femminile in difesa della Qabbalah, Rachel Morpurgo (1790–1871)," *Materia Giudaica* 15–16 (2010–11): 397–404.

65 Samuel David Luzzatto, *Viku'aḥ al Ḥokhmat ha-Kabbalah* (Gorizia, 1852), 53, 124.

66 Yosef Ḥayyim Brenner, *Shikkul ve-Kishalon* (Tel Aviv: A. Y. Shtibl, 1920), 90.

67 Shmuel Yosef Agnon, *Sefer Sofer ve-Sippur* (Jerusalem: Schocken, 1978), 334. See also S. J. Agnon, *The Bridal Canopy*, trans. I. M. Lask (New York: Schocken, 1937), 77, for another reference to *Sefer ha-Brit.*

68 Solomon Schechter, *Seminary Addresses and Other Papers* (Cincinnati, 1915), 1–2.

69 Dov Sadan, "Sefer ha-Bris," in *A vort bashteyt* (Tel Aviv: Varlag Y. L. Perets, 1978), 38–48, esp. 44.

70 Asher Ginzburg, *Iggrot Aḥad ha-Am* (Tel Aviv: Devir, 1923), 123.

71 Chaim Tchernowitz, *Pirke Ḥayyim* (New York: Hotsa'at Bitsaron, 1954), 79.

72 Sadan, "Sefer ha-Bris," 45.

73 Ibid., 46–48.

74 N. Naumov, "In a Dark Corner" (Hebrew), in *Ha-Eshkol*, translated from the Russian by S. Y. Posner (Cracow, 1902), 4:225–26.

75 Samuel Alexandrov, *Massekhet Nega'im* (Warsaw, 1886), 25.

76 Jeffrey Shandler, ed., *Awakening Lives: Autobiographies of Jewish Youth in Poland before the Holocaust* (New Haven: Yale University Press, 2002), 119. My thanks to Professor Kenneth Moss for this reference. For a book endorsed by so many orthodox authorities, the indignity of having to read it in a bathroom is a startling reminder that the traditionalists were never monolithic and that it is impossible to situate the many readers of *Sefer ha-Brit* in neat categories.

77 I am grateful to Professor Eliyahu Stern who shared with me his recent discovery, via the Yivo Library in New York, from the archives of the Rabbinical School and Teachers' Seminary, Vilna, 1847–1917 (RG 24), folder 70: a list of books submitted by the Rabbinical School to the Russian government on December 9, 1855. Entry 12 lists *Sefer ha-Brit* as an "encyclopedia of sciences, physics, natural history, popular astronomy, and theology. The text shows its quality in every part." The translation from the Yiddish is that of Professor Stern.

Sefer ha-Brit has even entered the discourse of Israeli politics. As recently as 2012, Chaim Amsalem, member of the Knesset, published a pamphlet in Jerusalem on the virtues of earning a living that cites Hurwitz's entire discourse on this subject.

78 Moshe Henekh Bernstein, *Oẓar Pitgamim Makhimim* (London, 1904), 17.

79 Issac Klein and Shalom Isaac Levitan, eds., *Sefer ha-Maggid* (Satu-Mare, Romania, 1906), Argaz Mikhtavim. Judah Leib Lazaravo, *Yalkut Yehudah* (New York, 1934), 40, repeats the same story in a Yiddish version.

80 Moses Joseph Shneerman, *Sefer Ohel Moshe al ha-Torah, Sefer Shemot* (Brooklyn, 2009), 46n.26. Cf. a different testimony on Rabbi Shach and *Sefer ha-Brit* in Eliezer Brodt, "Laws on the Blessing of Seeing [God's Creatures] in the Framework of the Book *Ma'agal Tov* of Ḥayyim David Azulai" (Hebrew), *Yeshurun Me'asef Torani* (New York: Makhon Yeshurun, 2012), 871.

81 Jacob Tessler, *Be-Aḥarit ha-Yamim* (Leeds, 1942, 2000), 151–54; the quotation is on 154.

EPILOGUE

1 This is not the place to offer a full bibliography, but I do refer to several other new works below.

2 On recent evaluations of Katz's work, see Jay Harris, ed., *The Pride of Jacob: Essays on Jacob Katz* (Cambridge, Mass.: Harvard University Press, 2002); and Israel Bartal and Shmuel Feiner, eds., *Historiographia be-Mivḥan: Iyyun Meḥudash be-Mishnato shel Yaakov Katz* (Jerusalem: Merkaz Zalman Shazar, 2008).

3 Jonathan Garb, "The Political Model in Modern Kabbalah: An Examination of the Writings of R. Moses Ḥayyim Luzzatto and their Intellectual Surroundings" (Hebrew), in *Avi be-'Ezri* (a volume honoring Professor Aviezer Ravitsky), ed. Benny Brown, Menahem Lorberbaum, Yedidia Stern, and Avinoam Rosenak (Jerusalem: Merkaz Zalman Shazar, 2014), 533–65; idem, "The Circle of Moshe Ḥayyim Luzzatto in Its Eighteenth-Century Context," *Eighteenth-Century Studies* 44 (2011): 189–202; and idem, "The Modernization of Kabbalah: A Case Study," *Modern Judaism* 30 (2010): 1–22; Mekubbal be-Lev ha-Se'arah: R. Moshe Ḥayyim Luzzatto (Tel Aviv: Tel Aviv University Press, 2014).

4 David Sorotzkin, *Orthodoxia u-Mishtar ha-Moderniyut: Hafakata shel ha-Masoret ha-Yehudit be-Eropah be-et ha-Ḥadasha* (Tel Aviv: Ha-Kibbutz ha-Me'uḥad, 2011); Maoz Kahana, "Bein Prague le-Pressburg: Ketivah Hilkhatit be-Olam Mishtaneh me *ha-Nodah be-Yehudah* ad *ha-Ḥatam Sofer*, 1730–1839," Ph.D. diss., Hebrew University, 2010; Ḥaviva Pediyah, "The Mystical Experience and the Religious World in Hasidism" (Hebrew), *Da'at* 55 (2005): 73–108; Boaz Huss, *Ke-Zohar ha-Raki'ah: Perakim be-Toledot Hitkablut ha-Zohar u-ve-Havnayat Arko ha-Simli* (Jerusalem: Makhon Ben Ẓevi, 2008); Shaul Magid, *From Metaphysics to Midrash: Myth, History, and the Interpretation of Scripture in Lurianic Kabbalah* (Bloomington: Indiana University Press, 2008); and Roni Weinstein, *Shibru Et ha-Kelim: Ha-Kabbalah ve-ha-Moderniyut ha-Yehudit* (Tel Aviv: Tel Aviv University Press,

2011). On Weinstein's book, see the critical remarks of David Malkiel, "Rapture and Rupture: Kabbalah and the Reformation of Early Modern Judaism," *Jewish Quarterly Review* 103 (2013): 107–121.

5 See Garb, "Political Model," 538. The latest study of Ḥai Riki is an unpublished paper of Saverio Campanini, "L'autobiografia di Immanuel Chay Ricchi, cabbalista itinerante." My thanks to Professor Campanini for allowing me to read the paper.

6 Eliyahu Stern, *The Genius: Elijah of Vilna and the Making of Modern Judaism* (New Haven: Yale University Press, 2013), esp. 2–11.

7 Ibid., 166–70; the quotation is on 168.

8 See, e.g., Moshe Idel, *Old Worlds, New Mirrors: On Jewish Mysticism and Twentieth-Century Thought* (Philadelphia: University of Pennsylvania Press, 2010), especially the introduction. Consider as well the heated response of Steven Aschheim, "New Thinkers, Old Stereotypes," *Jewish Review of Books* 9 (Spring 2012); available at http://jewishreviewofbooks.com/articles/179/new-thinkers-old-stereotypes. See also Yitzhak Melamed, "Solomon Maimon and the Failure of Modern Jewish Philosophy," and the discussion it engendered on the Association for the Philosophy of Judaism webpage: www.theapj.com/symposium-on-yitzhak-melameds-salomon-maimon-and-the-failure-of-modern-jewish-philosophy-2.

9 Gershon Hundert, "A Reconsideration of Jewish Modernity," in *Schöpferische Momente des europäischen Judentums in der frühen Neuzeit*, ed. Michael Graetz (Heidelberg: C. Winter, 2000), 321–32; and idem, *Jews in Poland-Lithuania in the Eighteenth Century: A Genealogy of Modernity* (Berkeley: University of California Press, 2004). Cf. also Ruderman, *Early Modern Jewry: A New Cultural History* (Princeton: Princeton University Press, 2010), 217–18.

10 I am indebted to a personal conversation with Elḥanan Reiner on this point. I cannot fully document this suggestion here, but consider, for the time being, on Spira, Agata Paluch, "The Ashkenazic Profile of Kabbalah: Aspects of the *Megalleh Amuqot* . . . by Nathan Neta Shapira of Krakow," *Kabbalah* 25 (2011): 109–30; on Eibeschütz, Pawel Maciejko, "Was Rabbi Jonathan Eibeschütz a Crypto-Christian?" paper presented at the 39th Annual Conference of the Association for Jewish Studies, Toronto, December 2007; on Ẓoref, Elḥanan Reiner, "To be a Sabbatean in the Eighteenth Century," paper presented at the Central European University, Budapest, January 21, 2013; on Naḥman, Zvi Mark, *The Scroll of Secrets: The Hidden Messianic Vision of R. Nachman of Braslav* (Brighton, Mass.: Academic Studies Press, 2010); on Lefin, Nancy Sinkoff, *Out of the Shtetl: Making the Jews Modern in the Polish Borderlands* (Providence, R.I.: Brown Judaic Series, 2004); on Menasseh Ilya, Isaac Barzilay, *Manasseh of Ilya: Precursor of Modernity among the Jews of Eastern Europe* (Jerusalem: Magnes Press, 1999); and on Meir Simḥah ha-Cohen, Louis Jacobs, "Rabbi Meir Simhah of Dvinsk," in *Judaism and Theology: Essays on the Jewish Religion* (London: Vallentine Mitchell, 2005), 196–208. See also Alan Brill, *Judaism and Other Religions: Models of Understanding* (New York: Palgrave Macmillan, 2010).

11 David B. Ruderman, *Jewish Enlightenment in an English Key: Anglo-Jewry's*

Construction of Modern Jewish Thought (Princeton: Princeton University Press, 2000), chap. 4, esp. 169–83.

12 Cf. Ruderman, *Early Modern Jewry: A New Cultural History*, chap. 1.

13 Olga Litvak, *Haskalah: The Romantic Movement in Judaism* (New Brunswick, N.J.: Rutgers University Press, 2012), 92–93.

14 Ibid., 12.

15 Ibid., 77: "The Jewish enlightenment is not the *Haskalah*, but the very idea is the work of the maskilic imagination, conceived by nineteenth-century Jewish romantics and championed by David Sorkin and Shmuel Feiner, their present-day intellectual heirs." For an excellent summary of romanticism that reinforces in many respects Litvak's understanding, see Warren Breckman, *European Romanticism: A Brief History with Documents* (Boston/New York: Bedford/St. Martin's, 2008), introductory essay.

16 Litvak, *Haskalah*, 32–33.

APPENDIX 3

1 This table of contents was prepared by Resianne Fontaine for a lecture she delivered in Amsterdam in February 2001. I thank Professor Fontaine for allowing me to borrow it. I have made minor emendations.

Index

Lefin, Mendel, 118
Leibniz, Gottfried, xii, 4, 7, 8, 9, 10, 11, 68, 69, 70, 72
Lemberg, 21–22, 32, 124, 125
Leopold, prince of Braunschweig-Lüneburg, 81–86, 156n49
Lessing, Gotthold Ephraim, 82–83, 84, 85
Levi, David, 119
Levin, Ẓevi Hirsch (Hart Lyon), 6, 23–24, 94, 113
Levinsohn, Isaac Beer, 63, 96–97
Levison, Mordechai Shnaber, 44, 73
lightning rods, 48, 105, 106, 147n26
Lindau, Barukh, 23, 24, 44, 90, 93
Linzig, Abraham, 38
Litvak, Olga, 115, 119–21
Locke, John, 73
Loewenstamm, Abraham, 79
Loewenstamm, Saul, 6, 20, 23, 24, 94
London, xi, 23, 26–28, 85, 113, 119
Love of neighbors. *See* moral cosmopolitanism
Lublin, 32, 53
Luria, Isaac, x, xii, 4, 27, 38–39, 40–43, 81, 105, 111, 116, 117, 123. *See also* Vital, Ḥayyim
Luzzatto, Moses Ḥayyim, 116–17, 120–21
Luzzatto, Samuel David, 63, 79–80, 111

Ma'aseh ha-Tartuffe, 15
Maggid, Shaul, 116
Maharal of Prague, 78
Maimon, Solomon, 4, 9, 23, 46, 67–72, 150n28
Maimonides, Moses, 4, 9, 13, 46, 47, 54, 64, 69, 72, 78, 94–95, 109
Mainz, 3
Malbim (Meir Leibush Michel Weiser), 159n23
Mark, Zvi, 105–6
Mas, Michael, 32
Matil (or Mottels), Itsik, 22, 141n15
Mazouz, Maẓli'aḥ, 111
McCaul, Alexander, 98, 108
Mecklenburg, Jacob Ẓevi, 79
Melamed, Abraham, 62–64
Meliẓ, Zelig, 32
Menahem Mendel of Shklov, 20
Menasseh ben Israel, 78
Mendelssohn, Moses, x, 4, 5, 19, 22, 23, 54, 79, 84, 91, 94, 97, 106, 117, 119, 120
messianism, 120: in *Sefer ha-Brit*, 36, 114
Mezerich, Judah Leib, 3, 6, 24, 25, 26
Michal, Yeḥiel, 79
Mieses, Judah Leib, 63, 95–96, 98
miracles. *See* nature, wonders and miracles of
Mitnagdim, xiii, 17
Mittleman, Allan, 71
Miẓvot Tovim (of Pinḥas Hurwitz), 38
monads, 7, 69, 70
moral cosmopolitanism: in Pinḥas Hurwitz's writing, 75–89, 98
Mordechai ben Nisan, 106
Morpurgo, Rachel, 111
Morteira, Saul Levi, 78
Moses, 13, 58–61, 63, 105; as politician and magician, 59–61
Munnikhuisen, J.H., 5
Münz, Moses, 6, 22, 25, 29, 30

Nachmanides (Moses ben Naḥman), 78
Naḥman of Bratslav, 96, 104–7, 118
Naḥmias, Eliyahu Levi ben, 107, 125, 127
Napoleonic Sanhedrin, 79
nature, wonders and miracles of, 10–14
Naumov, N., 112
Neumanns, Joseph Karl, 30, 123
Newton and Newtonianism, 27, 28
Nicolburg, 32
Nobel, Nehemiah Anton, 71

Oppenheim, Beer Ben Isaac, 29, 31, 32, 99, 123
orthodoxy, 57–58, 63, 73, 95, 121
Ovadiah Yosef, 103, 110, 111
Oven (Ofen=Buda), 6, 22, 24, 29, 30, 32

THE SAMUEL & ALTHEA STROUM LECTURES IN JEWISH STUDIES

The Yiddish Art Song
performed by Leon Lishner, basso, and Lazar Weiner, piano (stereophonic record album)

The Holocaust in Historical Perspective
Yehuda Bauer

Zakhor: Jewish History and Jewish Memory
Yosef Hayim Yerushalmi

Jewish Mysticism and Jewish Ethics
Joseph Dan

The Invention of Hebrew Prose: Modern Fiction and the Language of Realism
Robert Alter

Recent Archaeological Discoveries and Biblical Research
William G. Dever

Jewish Identity in the Modern World
Michael A. Meyer

I. L. Peretz and the Making of Modern Jewish Culture
Ruth R. Wisse

The Kiss of God: Spiritual and Mystical Death in Judaism
Michael Fishbane

Gender and Assimilation in Modern Jewish History: The Roles and Representation of Women
Paula E. Hyman

Portrait of American Jews: The Last Half of the 20th Century
Samuel C. Heilman

Judaism and Hellenism in Antiquity: Conflict or Confluence?
Lee I. Levine

Imagining Russian Jewry: Memory, History, Identity
Steven J. Zipperstein

Popular Culture and the Shaping of Holocaust Memory in America
Alan Mintz

Studying the Jewish Future
Calvin Goldscheider

Autobiographical Jews: Essays in Jewish Self-Fashioning
Michael Stanislawski

The Jewish Life Cycle: Rites of Passage from Biblical to Modern Times
Ivan Marcus

Make Yourself a Teacher: Rabbinic Tales of Mentors and Disciples
Susan Handelman

Writing in Tongues: Yiddish Translation in the Twentieth Century
by Anita Norich

Agnon's Moonstruck Lovers: The Song of Songs in Israeli Culture
by Ilana Pardes

A Best-Selling Hebrew Book of the Modern Era: The Book of the Covenant *of Pinḥas Hurwitz and Its Remarkable Legacy*
by David B. Ruderman

Lightning Source UK Ltd.
Milton Keynes UK
UKHW010650080920
369340UK00014B/159

9 780295 748061